The Ris
of the
Morning Star

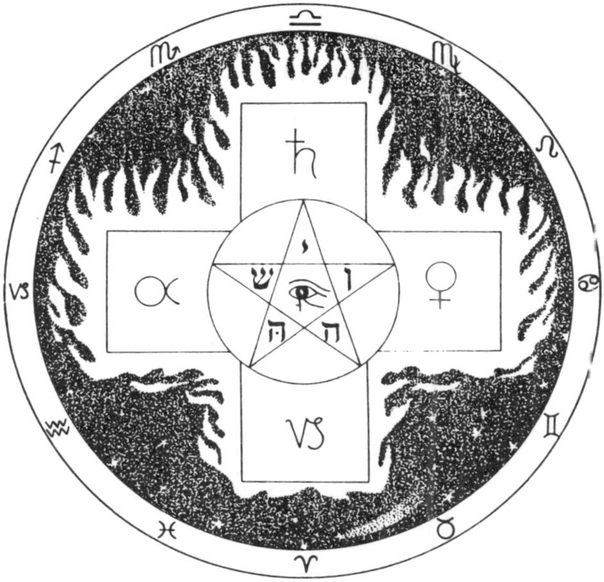

By the same Author;

The Path of the Dragon - Elemental Purification
Winged Serpents Dancing
Earth Link - A guide to the Planetary Awakening

Forthcoming;

Jewels from the Serpent's Nest - Understanding Male Sexual and Spiritual Reality

Vincent Selleck

The Rise of the Morning Star

The Return of the Love Goddess

An interpretation of the Myths and Mysteries of Venus/Aphrodite and their influence upon the resurgence of feminine spirituality.

SHEKINAH FOUNDATION
Byron Bay
Australia

First Edition 1990
Published by

Shekinah Foundation
PO Box 677
Byron Bay NSW 2481
Australia

Cover illustration - Tony Forbes.
Interior illustrations by the Author, Helen Gori and Eleni Mann.
Astrological charts by Astrologic - Byron Bay, Australia.
Typesetting and book design by Netlink Design.
Printed and bound in Australia by
Merino Lithographics, Moorooka, Queensland.

ISBN O 9587937 27

Dedicated to the Goddess in every woman.

May Her presence be revealed.

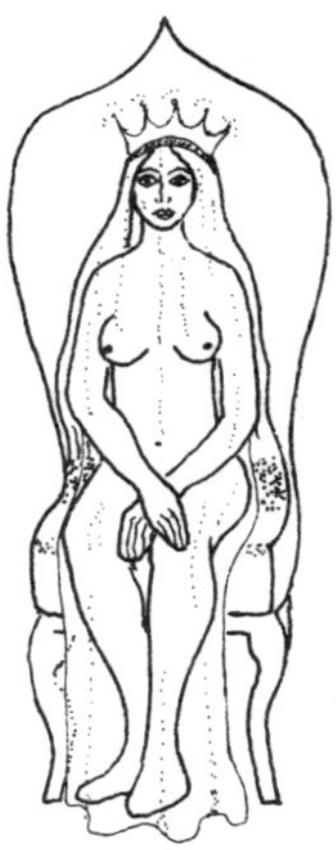

Acknowledgments

I thank the Goddess for the gift of Life and that She has revealed herself to me; that She has brought me to my knees to see that I am no better than any man. I thank Her for my tears and my pain; that I have known Love and great sorrow. I thank Her most for the embrace of Her warm shadow, for without Her darkness I would have never known what it was to see the Light.

This book would never have been written if it were not for the inspiration I gained from the many authors who preceeded me. Please take note of the titles listed in the bibliography and consider this my personal recommendation to read them. I make no apology that my work should resemble my predecessors in any way. They have all contributed to the writing and deserve credit.

My most sincere thanks and gratitude to Helen Gori, who took my life apart by revealing the Goddess to me. Without her patient support and careful tact in editing I am sure this book would have trod on more toes than it does now. Her perspective as a women has been vital to me in so many special ways.

Sincere thanks to Nic Jeanes for lending us his computer; to Richard Smykowsky for astrological support; Keith Stewart for saving the book with wizardry when the computer disc crashed and to Tina, Kavi, Jade and all the other Goddesses who have shown me love in this life. Sisters I am eternally grateful.

Sincerely,

The foolish bard of a thousand shattered dreams.

Contents

Preface

What a Fool I have been

What could a man possibly know about the Goddess and women's spirituality?

My fascination for the Goddess and Her mysteries came painfully to me when my tower collapsed. There I was, an initiate into the mysteries of the Father, the cosmos and Higher Intelligence, when suddenly, the vacuum of my separation from humanity imploded leaving me shattered on the ground. I thought I knew it all; the order of creation; the names of the angels; the codes of Higher Intelligence. I was so sure that I had all the answers; that I was chosen and on a special mission to help redeem the Earth. Now I feel like I have fallen from a very high place. It only took a direct experience of the Goddess for me to fall helplessly and willingly to my knees. How can such beauty exist? She is the siren and I must dash myself upon Her rocks.

Suddenly all the pain and suffering that I had blocked out of my life became unbearably real. If it were not for the incredible Love that accompanied the unbearable pain I would not have survived to write this book. Such is the paradox of Life. You cannot fly with your head in the clouds without having your feet firmly grounded on the Mother's soil.

The lesson I learnt was something I knew but had never fully experienced. There must be a balance between the Heavenly and the Earthly. The knowledge and energetic contact that I have experienced through the mysteries of the Father have now found balance with my presence on Earth. It is pointless to be obsessed with the afterlife when we are here and now in this body to serve the Spirit with our flesh. The Goddess is the redeemer of this world. Her presence and Spirit nurtures and protects the seeds of Life. The Earth is not corrupt and filthy, but potentially our paradise.

I can claim no authority for this work. If you find truth in these words, then well and good. If not then cast them aside, it does not matter. I do not expect to receive accolades from this humble exploration. I only hope not to offend too many, but some, unfortunately, cannot help but be offended.

Please forgive me my unworthiness.

A very insignificant human, on a very insignificant planet, in a very insignificant galaxy, in a vast and totally unimaginable universe.

Introduction

Brightly shines the Morning Star
the mystery that ends the night
towards Her flame my soul it flies
scorched like moth wings upon Her light

Vincent Selleck

For centuries, in cultures throughout the world, Venus, the morning and evening star, has captivated mystic attention and aroused sublime emotion. No other planetary body has inspired the depth of intrigue, ardent praise or vehment hate as has Venus. Her Goddesses and deities are numerous and extensive and include Aphrodite, Eos, Astarte, Al-uzza, Ishtar, Mylitta, Lucifer, Christ and Quetzalcoatl.

'The Rise of the Morning Star' examines Venus' mythic past; emergent, present and radiant future. She is awakening and restoring the beauty, strength and respect of the feminine to our male dominated world. Her spirit is the redeeming power that connects spirit and soul to our bodies; higher-selves to lower. Without an awakening of Her nurturning power and reverence for nature, Her kingdom, we will soon be custodians to a dust-bowl.

Within the pages of this book are revealed some of Her deepest mysteries. As you follow the ancient thread of Her rise, persecution and recent renewal you will discover keys to understanding the human dilemma. What is the true source of evil? What will happen to us when our solar system enters the Photon Belt? How can we experience spiritual and sexual ecstasy at the same time?

You will find, as I have, that Venus, the planet and Goddess, is the most esoteric, magical and mystical influence upon our lives. She inspires all art, music and magic. Venus is the only heavenly body, apart from the Moon and Sun, which is bright enough to cast a shadow. As She weaves a path between Sun and Earth, Her alignments create a pentagram against the stars. She guards the passage way between Light and Dark. As Morning Star She preceeds the dawn. As Evening Star She shines first at sunset.

I have reached into the myths and spiritual teachings of many cultures in preparing 'The Rise of the Morning Star', combining all of the pieces into a deep synthesis, full of revelation and supressed mystery. In writing this book I have been transformed, as each new fragment of the Truth has triggered a deep remembering and flooded my being with a rare Love and ecstasy.

I hope you will enjoy the journey as much as I have.

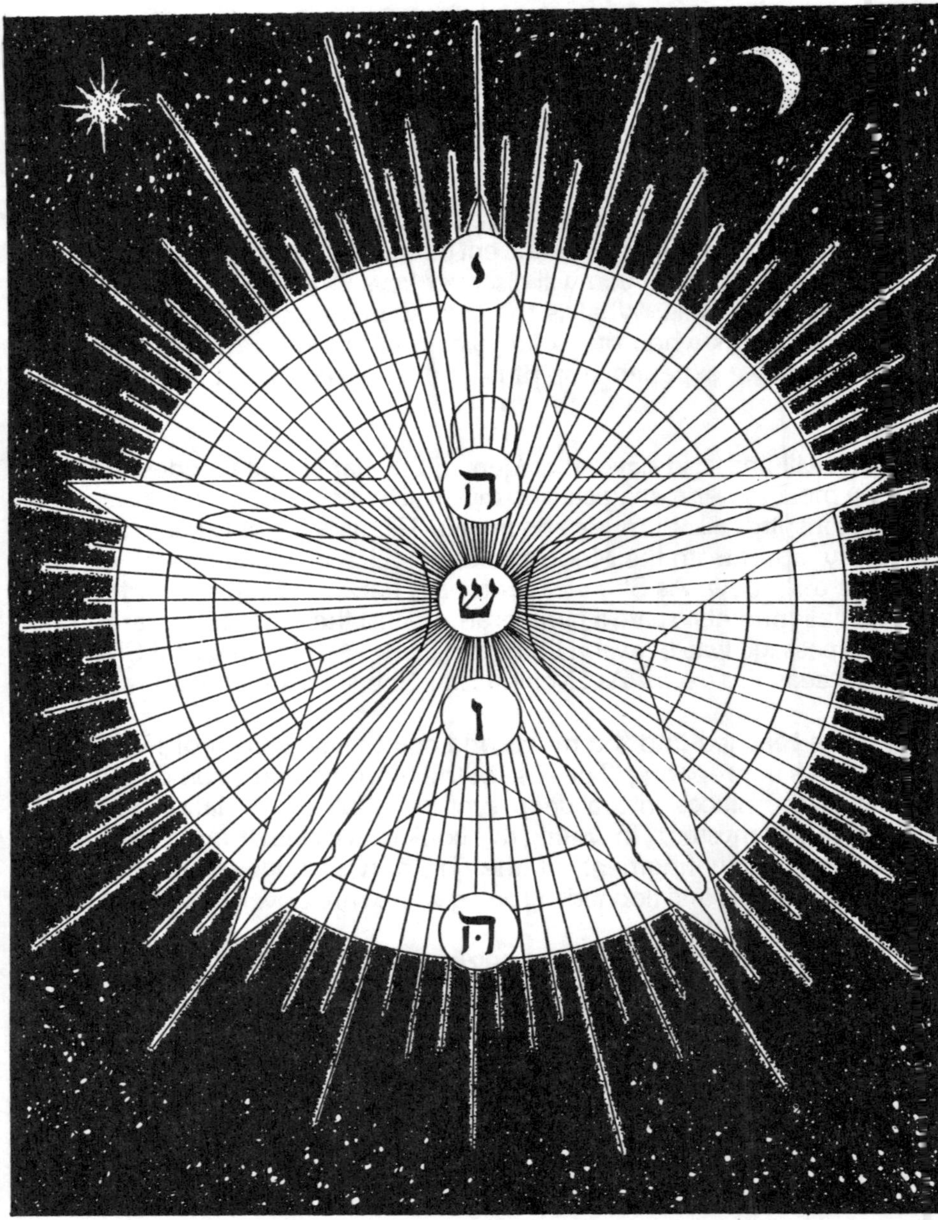
י
ה
ש
ו
ה

1

LOVE

The Redeeming Power of the Goddess

"The Virgin is believed to have died on August 13th, to have risen again and ascended to Heaven on the third day."

The White Goddess
Robert Graves
Pg 255

Our world is currently redefining the feminine. The over-emphasis on male virtues and vices that has resulted from the rise of the patriarchy has led man to positions of power throughout the world and has stifled femininity. Women have been viewed with derision and branded evil; a temptation to pure and holy males. Some see these attitudes as a reaction to ages of matriarchal rule, which existed before patriarchy, where the Goddess was triumphant as the principle source of life. Her dark side, the Death Crone, inspired fear among people. She was both creator and devourer. When males revolted and became owners of women, and therefore the life-giving process, all traces of Her worship were obliterated. As a result, six million people met their death through the inquisition and holy wars.

In today's 'civilised' world a new recognition of women and women's value is emerging. Centuries of persection are now coming to an end as women in some countries are becoming respected as equals to their sons and husbands. The roots of this change in attitude are being nurtured by a spiritual awakening, deep within the psyche. The Goddess is about to reveal herself as the redeemer of Life. She is the creator and Her Spirit is now awakening within each of us and within Life itself. Her nurturing of the psychic and emotional bodies and her caring for nature and Life are the principal elements needed to restore balance in the world. Her power is that of Love, which she inspires in each one of us and it is that Love which will carry us home.

The Myth of Creation

In Western societies the dominant creation myth is the Judeo/Christian account of Genesis. The One and only Father God creates the world and everything in it in seven days and only good exists. Woman is created as an after-thought to satisfy Adam's longing for a companion. The woman, Eve, then wants to know too much, is tempted by a serpent, and seduces Adam to eat from the Tree of Knowledge and all hell breaks loose. God said to woman; "I will greatly multiply thy sorrow" and "thy desire shall be to thy husband, and he shall rule over thee". Because of the temptress Eve's 'original sin' women have been considered spiritually inferior and in many cultures this attitude still persists today. Our only path of redemption is through the male Son to reach the Kingdom of the male Father after the Earth has been obliterated by males in the war 'Armageddon'. As stated in Revelations 14:4, the Saints of the new world are "the 144,000 of the Lamb" who "did not defile themselves with women; in fact, they were virgins" and "without blemish".

This creation myth is one of the foundation stones of the power elite of the Western world. Fundamentalists like Ronald Reagan believe that the Earth is going to be consumed by fire, nuclear war, and that this is the rapture that precedes entry into the Kingdom of Jesus. While this account of creation (and the subsequent destruction) of the world may satisfy the ninety million fundamental Christians in America, other creation myths exist which are just as valid to their proponents. In particular the Gnostic story of creation, as interpreted by the Mother (Mirra Alfassa) from the work of nineteenth century Gnostic Max Theon, deserves closer consideration.

The foundation of this Gnostic cosmology is the principle of emanation. The absolute supreme source of everything breathes out its own essence to create other beings or Gods which also breathe or emanate other beings, worlds and the like. Thus a chain of causality is constructed through different levels of emanation, all connecting to the original source.

One such emanated Goddess is cited in the following passage with the creation of the Cosmos.

> "So She made her first four emanations (to serve as powers for the creation of the Cosmos). The first was Consciousness and Light ...; the second was Ananda (bliss) and Love; the third was Life; and Truth was the fourth. Then, so the story goes, conscious of their infinite power, instead of keeping their connection with the Supreme Mother and through Her, with the Supreme, instead of receiving indications for action from Him and doing things in proper order, they were conscious of their own power, and each one took off independently to do as he pleased ... They forgot their origins. And because of this initial oblivion, consciousness became unconsciousness, and Light became darkness; Ananda became suffering, Love became hate; Life became death; and Truth became falsehood. And they were instantly thrown headlong into what became matter".

'*Mother's Agenda*', Vol. 3,
1962, pg 55.

The above indicates that creation on the material plane was inherently evil through cosmic error or design. Is humanity therefore blameless for the inherent darkness within our lives and merely subject to the forces of primal creation themselves? I say cosmic error or design because, from the point of view of the Supreme or First Cause being perfect, and therefore (by a certain way of thinking) good, the creation of evil must have been an error. Yet without evil, darkness and suffering there would be no Light or Love. Adversity is the testing ground of the soul and the wealth of living experience. Hence spiritual evolution is therefore only gained through trial, hardship and eventual victory over the elements of darkness.

In this way sin is seen to be original, not solely to humanity, but to the entire cosmos. Did the grand mind of minds therefore dream the vision of Light **and** darkness into the original fabric of space and time as an evolutionary condition for testing spiritual creations?

Returning to Mirra Alfassa's fallen emanations, the Supreme Being found it necessary to redeem creation:

> "And once the world has become like that; has become the vital world in all its darkness, and they have created Matter, the Supreme Mother sees the result of her first four emanations and she turns towards the Supreme in a great entreaty: 'Now that this world is in such a dreadful state, it has to be saved'! We can't leave it in this way, can we? The divine consciousness must be given back to it. What to do'? And the Supreme says 'Thrust yourself into a new emanation, an emanation of the ESSENCE of Love, down into the most material Matter'. That meant plunging into Matter, and that became the primordial source of the Divine within material substance. And from there she begins to act as a leaven in matter, raising it up from within".
>
> Ibid.

This explains the innate femininity of matter, nature and the Earth. The emanation of the Divine Mother-Spirit has impregnated every atom of substance, endowing it with the essence of Love. The Roman Church recognizes this as the Holy Ghost or Spirit, the third aspect of divinity. The Hebraic Kaballah calls this Holy Spirit the 'Shekinah' or bride of God. Yet this essence has been lost, like the tribes of Israel and the Holy Grail. When it is restored it is said that the Messiah shall return.

So where shall we look for this lost Shekinah? In the billions of stars and planets of the infinite sky? In our quest for might, power and authority? In material wealth, comfort and the good things of Life?

The infinite redemptive Spirit is with us now, in every atom of our beings. The Mother and I are one. Within the unexplored aspects of our feminine natures, the unconscious mind or dream-mind, lies the redeeming Holy Spirit. The Holy rite of the Shekinah is surrender; immersion of the self into the sea of primal substance, the Light-form which permeates and is the substance of all creation. In the inner mystical sea of Light and energy the male mind, the logical, reductive, separatist self drowns in a sea of surrender. This is the fear which men have. The Goddess will swallow them all before they reach heaven, nirvana or the blessed isle. For this she is despised most of all.

The soul, the immortal God spark of evolving, continuing consciousness knows the Mother and loves and respects Her ways. The darkness of primal creation is the testing ground for the soul. The immortal spark of divine intelligence realises itself within the continual contrast of the eternal dark. As experience teaches the soul more about its truth; Love and inherent beauty, the feminine Spirit is discovered within. Soul and Spirit merge into one dance and are lifted to an arena of clear perception and realization of Truth. The Mother, or indwelling presence, educates the soul about the Light which shines in the heart of darkness and also the darkness in the heart of Light. The Taoists have represented this Truth in the symbol of the Tao.

In the *'I Ching'* or Book of Changes, Light and Dark, Yang and Yin, creative and receptive, are not seen as inherently hostile towards each other. Rather, they combine and interact to create the rhythm of the seasons and the cycles of nature.

The apparent good and evil of the world are aspects of the same creative impulse. The alchemical proving ground for this realisation is the inner world of our own darkness. Our fear exists only as unexpressed joy. Our hate is the Love that we do not feel for ourselves. Overbearing authority is a disguise for inability to surrender. All these qualities and feelings exist as contradictory higher and lower aspects of our own beings. When we see ourselves; our souls, longing for expression, the simple blindness that separates inner and outer dissolves leaving only Truth. The Truth of hate is Love. The Truth sees only one being in all things. This is neither good nor evil, it simply is.

The key to breaking down the resistance to Truth is, for our present world, the cultivation of the feminine dream-mind. In order to see the Truth of One, both polarities need to be present. Imbalance on the side of the male mind has resulted in an inability to create beauty and equilibrium. The path between worlds cannot be entered while consciousness remains trapped in the one pole.

I have read many accounts of the atrocity and injustice that the male mind, working through patriarchy, has spawned. While I see the feminist and pagan perspective clearly and agree with them, I also see that Truth does not favour either the male or female mind. Truth recognises only the Love which unites them and makes them whole. A rebalancing is necessary. The repression and persecution of the feminine dream-mind and its manifestation needs to end. The nurturing characteristics of the Mother could easily be predominant for some years without ill effect, but only unity of both minds can save us, our souls, and our planet in the end.

My purpose in this work is to bring to Light that which has been neglected. The Goddess is re-emerging within us all, to answer the needs of the time; to bring peace and Love to the Earth that shines. Her dance is a mating dance. Her power is the Love which draws her husband to Her and causes him to court Her with tenderness and respect. He will do anything to please Her. His wilfulness and rigid authority are put aside once the sweet fragrance of Her embrace draws him to submission. Irrational, inconceivable, unexpected and totally intoxicating Love will bring him to his knees. This is the Goddess released. This is the world redeemed.

2

VENUS

The Suppressed Mysteries

> *"The curse in Genesis on the woman, that she should be at enmity with the serpent, is obviously misplaced: it must refer to the ancient rivalry decreed between the sacred king Adam and the Serpent for the favours of the Goddess ..."*
>
> *The White Goddess*
> *Robert Graves. Page 257*

The occult, secret and mysterious nature of the planet Venus is evident in the myths and spiritual teachings of many cultures. In the South American legends of the Aztec Indians, Quetzalcoatl, Avatar/God of wisdom and 'Place where the Serpent Dwells', is identified with the morning and evening stars. Numerous Goddesses are attributed to Venus including Ishtar, Al-Uzza, Astarte, Eos, Aphrodite and Isis. In Egypt, Venus was called Muth, which means 'Mother of all Living'. Even Christ is connected with the mystery of Venus, the Morning Star:

> "I Jesus have sent mine angel to testify unto you these things in the churches. I am the root and offspring of David, and the bright and morning star".
>
> '*Revelations*' 22:16

The string of connections between Saints, Avatars, Gods and Goddesses with Venus gives this planet a paramount role in spiritual affairs upon the Earth. Madame Blavatsky in her *'Secret Doctrine'* says of Venus:

> "Venus is the most occult, powerful and mysterious of all planets; the one whose influence upon, and relation to, the Earth is most prominent".
>
> *'Secret Doctrine' III*
> Page 42

With such a wealth of mystical tribute being paid to the Earth's closest neighbour it is a wonder that more attention is not being paid to Venus in astrology and spiritual science. Our dear, sweet sister Venus guides our path of redemption more closely than any of the other planets or luminaries. Though She is considered to be our 'twin sister' and 'the Earth's primary or spiritual prototype' (*S.D. III*, page 33, 35) a complete veil surrounds Her like the clouds that eternally cover Her face.

Venus - Lucifer

Perhaps the greatest reason for the Venusian cover-up is Her relationship to Lucifer, the fallen angel or 'shining one'. Manly P Hall in his *'Secret Teachings of All Ages'* explains in part this connection:

> "Being visible in the sky at sunset, it (Venus) was called Vesper, and as it arose before the sun it was called the false light, the star of the morning, or Lucifer, which means the light-bearer".
>
> (pg LXVIII)

For the average Christian the name Lucifer conjures up visions of evil that are synonymous with the devil or Satan. This was not always the case however. St Peter in *'Des Esprits'* considers Lucifer in a more favourable light:

> "May the divine Lucifer arise in your hearts".
>
> *'Des Esprit', Tome IV,*
> Page 34-8

Madam Blavatsky also exposes the early Christian favour that Lucifer enjoyed:

"Venus, or Lucifer - also Shukra and Ushanas - the planet, is the light-bearer to our Earth, in both the physical and mystical sense. The Christians knew it well in early times, since one of the earliest Popes of Rome is known by his pontiff name as Lucifer".

S.D. III, page 45

Much dispute exists between pre-Christian Earth religions and the Church of Rome over the blanket denial of earlier deities by the church.

Traditions of ritual and worship that spanned centuries suddenly became profane and 'pagan' under the rising power of Christianity. The Goddess religions in particular, such as Isis and Aphrodite, became outlawed where the Church gained political and economic control. Our historians refer to these times as the 'Dark Ages' because the Christian patriarchy would prefer to keep these times of persecution and religiously condoned murder in the dark. So too, the truth about Lucifer is kept in the dark.

Lucifer - first Son of the Goddess

Archeology demonstrates that the first God-forms were female. The earliest images of art discovered were the 'Venuses' of the Paleolithic era (35,000-10,000 BC). These full-bodied, nurturing images represent the mysterious power of the female to create life within. It was thus women who symbolised the mystery of creation and the source of life in early history and pre-history.

The advent of male deity did not occur until much later:

"In Europe there were at first no male Gods contemporary with the Goddess to challenge her prestige or power, but she had a lover who was alternatively the beneficient Serpent of Wisdom, and beneficient Star of Life, her son".

Later in the same paragraph:

"The son, who was also called Lucifer or Phosphorus ('Bringer of Light') because as the evening star he led in the light of the Moon, was reborn every year, grew up as the year advanced, destroyed the serpent, and won the Goddess' love".

'The White Goddess'
Robert Graves
Page 387

When the patriarchy assumed control through the church, they could not abide with the idea of Lucifer being subservient to his Mother; even though his power and ancient influence upon the human spirit far exceeded the newcomer Jehovah. Yet he could not be denied, so Lucifer became Satan the devil and was despised by the Church of Rome.

Madame Blavatsky agrees:

> "There is a whole philosophy of dogmatic craft in the reason why the first archangel, who sprang from the depths of chaos, was called Lux (Lucifer), the "Luminous Son of the Morning", or Manvantaric Dawn. He had been transformed by the Church into Lucifer or Satan, because he is higher and older than Jehovah, and had to be sacrificed to the new Dogma".
>
> *'Secret Doctrine'*
> page 138

The Fall of Lucifer

It is plain to see that the disrepute that befell Lucifer 'Son of God' (*Job* 1:6) was largely the construct of the church. The true 'fall' of Lucifer was his arrival to Earth in the first place. Lucifer according to *'Isaiah'* was the 'Son of the Morning' who had 'fallen from heaven'. Yet his descent into the emanated world was out of service rather than as a punishment. It is only a recent device of the church which discredits him.

> "In the earliest days of the religion from which the Christian beliefs have sprung, Satan was worshipped. He was understood to be the greatest benefactor of man, placing before him temptations and experience without which he would remain stagnant and futile".
>
> *'The Finding of the Third Eye'*
> Vera Stanley Alder
> pg. 144

Referring again to the nature of darkness given by the Mother in her *'Agenda'* quoted in chapter one of this book, it can be seen that darkness only serves Light, for the purpose of educating the soul. Lucifer and the other adversaries who were once in favour but are now denied can be likened to the original emanations of

Love, Light, Ananda (bliss) and Truth which were corrupted by pride and thrown headlong into matter.

The Christian purgatory for Lucifer of endless torment in the pit of Hell is somewhat of a harsh treatment for a 'Son of God'. In the Hindu version, Lucifer, or the Mahasura, is given opportunity to repent. Through devotion, Love and service the fallen Lucifer can once more reach the status of being One with the deity. Even Lucifer's brother, the Archangel Michael does not presume to punish his errant brother:

> "But when Michael the Archangel had a difference with the Devil (Lucifer) and was disputing about Moses' body, he did not dare bring judgement against him in abusive terms, but said; "May Jehovah rebuke you!"
>
> *'Jude'* 1:9

If Michael, sword-bearer and heavenly protector, commander of the hosts of God's angels 'did not dare bring a judgement against him', who are we to do so?

Each of us must look into our own souls for the answer, for each of us must face the adversary within. Only by overcoming our own darkness are we free to see Lucifer in the clear Light. Though he is cast before us as an obstacle, his purpose is to test us and give us strength. Lucifer guards the way to enlightenment and it is he who tests us before we may pass into the kingdom of Heaven, just as Jesus was tested in the wilderness:

> "Now Jesus, full of the Holy Spirit, turned away from the Jordan, and was led about by the spirit in the wilderness for forty days, while being tempted by the devil".
>
> *'Luke'* 4:1-2

Here then is the key to the secret relationship between Jesus, Lucifer, Venus and the Holy Spirit. The Holy Spirit, the redeeming emanation of the Great Mother, led Jesus into the wilderness to be tempted. He could not be free to enter the Kingdom of Heaven until he had passed the test of his brother Lucifer. Lucifer and Jesus were thus united in the purpose of the Holy Spirit which we will see is the principle emanation of the planet-Being Venus.

In truth then, we are the judges of angels. The hosts of Lucifer/ Satan by design and in accord with the order of God, serve as temptors for the evolving souls on Earth. Their purpose is as much guided by the Holy Spirit as the mission of Jesus. Yet we must not fall for their cunning temptations, but discover the true Light of our inner heart which knows the way of Heaven. As we are told in *'First Corinthians'*:

> "Do you not know that we shall judge angels..."
>
> *'1 Cor'* 6:3

Clearly then, those who seek to have Lucifer eternally bound and tormented are those who are least able to withstand his temptations. When the redeeming influence of the Holy Spirit is present in the heart, Love overcomes selfish desire and no temptation is possible. The other face of Lucifer is his true face as an Angel of Light working with Christ for the betterment and perfection of humanity. **Our test on this level of creation is to realize the source of Light that exists in the heart of darkness.**

Let me close on this subject with the words of David Spangler:

> "Lucifer prepares man (woman) in all ways for the experience of Christhood and the Christ prepares man (woman) for the experience of God. Jesus said, 'As the Christ, I am the way the truth and the life. No man, goes to the Father but through me'. This is true. The avenue out of microcosmic limitation into macrocosmic wholeness, universal consciousness and attunement is through the Christ. But the Light that reveals to us the presence of the Christ, the Light that reveals to us the path to the Christ comes from Lucifer. He is the Light giver. He is aptly named the Morning Star because it is his light that heralds for man (woman) the dawn of greater consciousness. He is present when that dawn is realized. He stands no longer as the tester or the tempter but as the great initiator, the one who hands the soul over to the Christ and from the Christ on into even greater realms".
>
> *Reflections on the Christ*
> page 49-50

Serpent/Son of the Morning Star

Not only is Venus connected to Lucifer, but also to the Serpent, another symbol which has brought derision upon Venus and Her worship. According to an early creation myth the original Goddess who created the world had a lover who was alternately the Serpent of Wisdom and the Son or Star of Life. The Love of the Goddess was too intense for the Son and he was destroyed, but from his ashes arose a serpent. Both were cyclically destroyed and reborn to mate again with the Goddess.

The continuous alternate re-creation of the son and the serpent parallels the Venusian planetary cycle. Venus is alternatively the evening star, visible in the western night sky just after sunset and the morning star visible in the eastern sky before dawn. The Goddess is represented by the Sun who consumes her lover with Her burning passions. When Venus is transiting before or behind the sun, the planet becomes invisible for about twelve days. This disappearance parallels the mythological destruction of the serpent and the son in turn. Venus has therefore become a symbol for the regeneration of cycles.

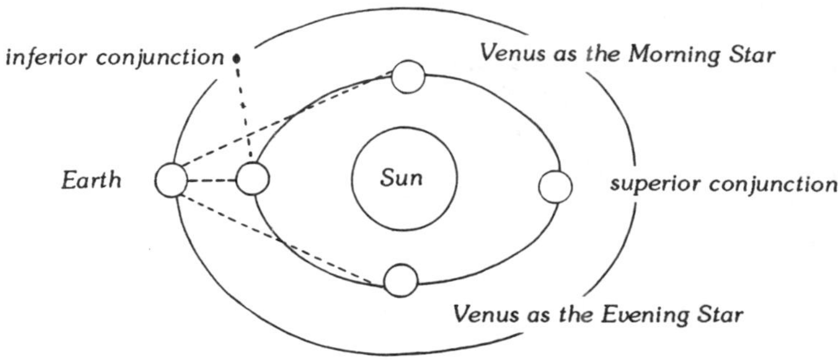

The alternate birth and death of the dual Son/Serpent is also paralleled in the mythologies of the Aztecs and Mayans regarding Quetzalcoatl. As the Evening Star, Quetzalcoatl is known as Tezcatlipoca or 'Smoking Mirror', the revealer of darkness. In his more familiar guise he is Quetzalcoatl, Morning Star, Lord of the Dawn, bringer of peace.

One of the main reasons for the suppression of Venusian based mythologies is this connection of light and darkness within the same mythic vehicle. The patriarchy has made stringent efforts to divide good and evil, light and darkness and man and woman.

Religions that respect the complimentary nature and mutual attraction of these opposites have been systematically persecuted as a result. None more so than Quetzalcoatl at the hands of the conquistadors and the inquisition. The greatest blasphemy, it seems, is to tell the truth!

Christ and Lucifer - the Son and the Serpent

The shadow, or Serpent aspect of Quetzalcoatl is also shared by the Christ in the mythic revelation that Lucifer is his Brother-self. Quetzalcoatl's descent into the underworld parallels Jesus' temptation in the wilderness. His brother Satan (Lucifer) offers him power to rule the world as king but he refuses. The apparent separation that has occurred between Satan, (the Serpent) and his brother-self Christ is due to the obscuration of their connecting link, the Goddess who births them both cyclically. The Goddess in this case is the Holy Spirit which moves their lives as two opposite components of a universal play. Both Christ and Lucifer are wedded to Her purpose, but each has a different role to play. Jesus, as the Son and Morning Star is the light-giver while Satan (Lucifer) is the Serpent and dark temptor of the soul. Together they are one mythic complex, engineered by the Holy Spirit to test and redeem human souls.

The serpent is perhaps one of the most universally applied symbols in human mythology. As Lucifer (Satan), the serpent is earth-bound and intent on drawing humanity into the darkness of selfish desires. As the Winged Serpent, Plumed Serpent, or Dragon of Wisdom, as in Christ and Quetzalcoatl, the serpent takes on an image of transcendence and enlightenment. Both of these images represent different aspects of the same mythic concept, but in different cycles of manifestation, either as morning or evening stars.

The essential energy of the serpent is the power of generation; the Earth 'Shakti' that generates life. These forces can be liberated through such practices as 'raising the serpent', the activation of Kundalini (spiritual fire) within the bio-psychic force centres or chakras. The recipient of this activation experiences ecstatic states of spiritual pleasure. Alternatively the energies of the serpent can be focused in the acts of generation, the continual focus of mind, soul and body on the physical part of creation and sexual reproduction. The advent of Western materialism as the primary culture on the planet represents the triumph of the dark serpent over the affairs of humanity.

Essentially, both directions of expressing serpent energy utilize the same force, but differ in motivation and focus. The energy of their unifying element, the Goddess or Holy Spirit, is needed to balance the continuous struggle between the complimentary elements of the human condition. When the Holy Spirit is present, the opposite sides of Being are seen to be essential to the whole. The acceptance of the shadow or dark aspect of our natures is thus a prerequisite to becoming unified within the Self. Good never totally destroys evil, but rather a balance or truce is achieved.

The darkness is no more than the part of ourselves that we deny, just as the Goddess has been denied. Our inner shadow remains self-destructive until it is loved and accepted. This means loving aspects of ourselves that we have judged to be unacceptable. The unloved and hidden aspects of ourselves manifest as an external evil, brought about by powerful subconscious projections from the denied aspects of our own beings. Without proper expression, the darkness within remains cloaked in guilt and shame and this reinforces the negative imprint of self. We are taught to fear our own darkness, rather than exploring our potential for self-integration and acceptance.

The Goddess has been denied and cast out as a dark shadow, yet it is Her magnetism and connectiveness that binds our hearts and heads together. She is the creative source needed to unify the Son and Serpent within. Her love binds the warring elements together in a greater purpose, unity and power. The combining of vital light and dark essences within the individual integrates both male and female polarities. Men can thus relate to their feelings and women to their intellects. The underlying message inherent in the myths and mysteries of Venus is the attainment of inner balance. Her love unites both light and dark as essential and complimentary aspects of an integrated whole.

Winged Serpents

The Winged Serpents or Sons of Light that have manifested upon the Earth can be identified by their associated symbols. Hermes, the wing-footed messenger of Olympus, carries the Caduceus or 'Winged Serpent' staff. In his Egyptian role as Thoth, the Son of Light, he vanquishes Typhon, the dragon of ignorance and perversion, by striking the beast with the magical 'Winged Serpent' staff.

The Cadaceus itself is symbolic of the cycle of Venus. The two serpents wind alternately around the staff which bears aloft the winged orb, an Egyptian hieroglyph that represents the Sun.

Clearly this indicates the alternate manifestation of Serpent and Son, evening and morning stars around the central orb of the Sun. Apollo is also linked with the dragon or serpent and the Sun He, like Thoth, strikes the serpent, Python's head. So too Krishna slays the black serpent Kalinaga. Thor, thunder God of Scandinavia, also strikes the serpent in the head with his mace of crucifixion. All slain dragons and knights such as St George are connected to Venus, the root and source of world mystery. An entire chapter of *'Revelations'* is devoted to re-telling the story of the Goddess and Her son the serpent:

> "And there appeared a great wonder in heaven; a woman clothed with the sun, and the moon under her feet, and upon her head a crown of twelve stars":
>
> *'Revelations'* 12:1

'In heaven' means in the sky, and the 'woman clothed with the sun' refers to the spiritual presence of the Goddess as the Sun. The 'twelve stars' of her crown are the twelve constellations of the zodiac through which the Sun passes in one year.

> "And she being with child cried, travailing in birth and pained to be delivered".
>
> *'Revelations'* 12:2

Her son's birth, or the Morning Star aspect of Venus is imminent.

> "And there appeared another wonder in heaven; and behold a great red dragon..."
>
> *'Revelations'* 12:3

The serpent aspect.

> "... and the dragon stood before the woman which was ready to be delivered, for to devour her child as soon as it was born".
>
> *'Revelations'* 12:4

The rivalry between son and serpent or dragon is clearly evident as one tries to destroy the other.

> "And she brought forth a man child, who was to rule all nations with a rod of iron: and her child was caught up unto God, and to his throne".
>
> *'Revelations'* 12:5

The child emerges and is 'caught up' into 'God's throne'. The child is placed in heaven as the Morning Star, the Christ principle.

> "And there was war in heaven. Michael and his angels fought against the dragon; and the dragon fought with his angels, and prevailed not; neither was their place found any more in heaven".
>
> *'Revelations'* 12:7-8

The Son aspect represented by Michael, defeats the dragon or evening star and its place is no longer in heaven. The evening star thus disappears as the morning star is born.

The Snake-and-Bird Goddess

This mystic exploration of the serpent and Goddess would not be complete without revealing the nature of the Serpent-and-Bird Goddess. She was worshipped in central and South Eastern Europe from about 6500 B.C. Her triple form is represented in (1) the serpent - the latent psychic power of the reptilian dream-mind (2) the tree - representing the spine and (3) the soaring bird - of enlightened consciousness. The serpent powers, in the East known as Kundalini, climb up the tree or spine giving rise to spiritual ecstasy, the winged bird of self transcendence.

The mythic treasure in the symbol of the snake and bird Goddess is shared by many cultures. The South American Avatar Quetzalcoatl takes his name from the symbol. 'Quetzal' is a brightly coloured bird and 'Coatl' means serpent. Quetzalcoatl instructed the Aztec's and, later as Kulculcan, the Mayans in the arts of wisdom or teaching the snake how to rise up the trunk to become a bird or plumed serpent.

The Goddess can also be recognized as Eve in *'Genesis'*. By eating the fruit of the forbidden tree she is offered eternal life by the serpent who says:

> "For God doth know that in the day ye eat thereof, then your eyes shall be opened, and ye shall be as Gods, knowing good and evil".
>
> *'Genesis'* 3:5

Eve, as a woman, was first able to transcend mundane consciousness before Adam because of a little known neurological peculiarity.

> "According to some researchers, the corpus collosum in the female brain contains an abundance of nerve connections linking the right and left hemispheres; they found this not to be true of the male brain. Thus there is some neurophysiological evidence that males do tend to separate and compartmentalize 'feeling' and 'logic', while females tend to synthesize them".
>
> *'Science Digest'*
> No. 11, Nov. 83
> page 85

I am sure that with the correct development of the Anima, or feminine within the male, that men too can connect the twelve mid-brain pairs so that spiritual and sexual ecstasy can be experienced at once. Women, however, seem to have this capacity innately. The linking of the left and right hemispheres by the neo-cortex allows a synthesis of energy to flow between male and female aspects of the mind. This bi-sexual or androgynous exchange is the equivalent of making spiritual love to oneself, precipitating in powerful spiritual enlightenment.

The Mayans knew about the three-fold nature of the brain which they symbolize in the following three figures.

Chicchan - Reptilian brain

Oc - Mammalian brain

Eb - Human brain

In the Mayan symbols for the three parts of the brain, OC, or mammalian brain, relates to the right brain, the feminine, feeling and merging aspect of the mind. EB, the human brain relates to the left or logical, rational mind. In these two aspects humanity has been polarised as male and female - thinking and feeling, forgetting the roots of consciousness, the reptilian mind. Until it is awakened, according to Yogis of the East, the serpent force of Kundalini remains dormant at the base of the spine. Once activated, however, the serpent power rises to the crown chakra, on top of the head. This is also the area in which the mid-brain pairs are located. Once the neo-cortex is active all three brains are working together bringing about spiritual and physical harmony in thought, feeling and action.

Women's innate capacity to activate the neo-cortex and 'raise the serpent' set her apart in early times as the principle sex involved in mystic and magical arts. In Africa this led to the worship of the Black Goddess, who was bisexual, and thereby able to fertilise her own ovum. She was the 'witch who carried a snake in her belly'. Many primitive cultures did not realize that the male had anything to do with procreation. The inserting of the penis into the vagina was seen as only helping to open the womb so the spirit might enter.

The Golden Flower

The Taoist religion in China is one of the best preserved matriarchal cultures on the planet today. The Taoist *'Secret of the Golden Flower'* preserves the art of connecting the twelve mid-brain pairs in a symbolic breath meditation. The principle of the utilization of rhythmic breath is well understood by most major esoteric paths. In the 'Golden Flower' the breath is regulated into a circular orbit within the body with six counts or pulse beats for the exhale and six counts for the inhale.

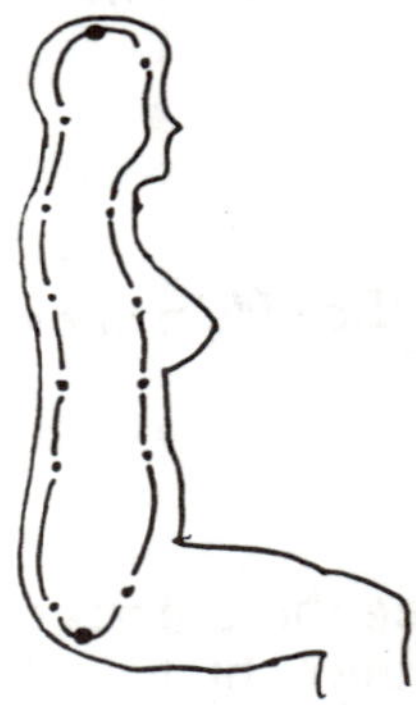

The Microcosmic Orbit

This subtle balance of the polarities of in and out breath through a twelve pulse beat cycle is a symbolic means of linking the twelve mid-brain pairs. The breathing path connects the serpent's resting place at the base of the spine directly with the serpent's flying place at the crown of the head.

The 'Golden Flower' breathing practice is also utilized in Taoist sexual practices. Since Taoism is essentially a matriarchal philosophy there is evidence to suggest that women initiated their men in the practice of 'drawing nectar up to the Golden Flower', activating the Kundalini, uniting the two brains, and teaching the serpent to fly through sexual union. The aim of the exercise of uniting the rhythms of breath together in love making is to experience sexual and spiritual ecstasy **at the same time.** Through the practice of connecting the orbital breaths of the male and female in lovemaking the undeveloped neurological pathways in the male brain are connected and the neocortex in the female activated.

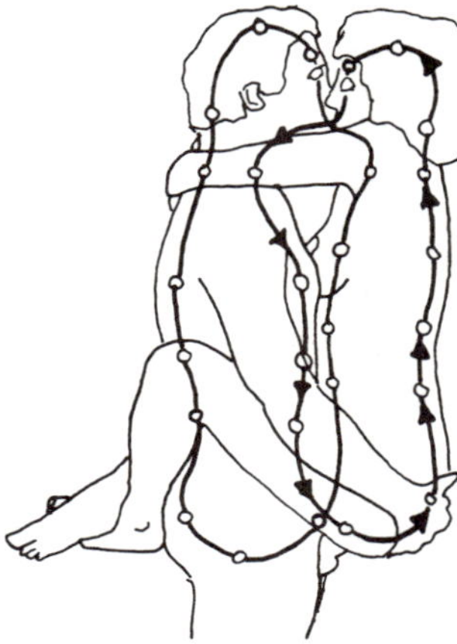

Fusion of microcosmic orbits in lovemaking

Readers wishing to engage in a deeper study of these techniques are directed to the *'Healing Love through the Tao'* series by Mantak and Maneewan Chia as listed in the bibliography.

The roots of sexual magic were the realm of women, as priestesses to the Goddess. Sex was not primarily a means of procreation in humans, but a magical means for spiritual transcendence and ecstasy. The evolutionary miracle of the menstrual cycle overcame the animal oestrus cycle in women. While men did not vary significantly in anatomical function from other primates, females developed the clitoris and the menstrual cycle which made sex pleasurable and possible all the time. The whole development of humanity as a spiritual, thinking creature is probably owed to woman's capacity to experience spiritual and sexual pleasure together in lovemaking.

In many of the Goddess religions of pre-Christian and pre-patriarchal times, priestesses would initiate men sexually in the temples. In some cultures, as in the Temples of Mylitta in Babylon in the fifth century B.C., once in their life each Babylonian woman would give her self in the temple to a man who so chose her. This was her duty to the Goddess. Similar services were offered in the temples of Astarte, Ishtar and Aphrodite. This practice is not to be condemned as mere prostitution, but rather it must be understood that women in those times were not subject to the alienating and psychologically damaging influences of todays repressive patriarchy. Women's mysteries, foremost of these the sexual and spiritual mysteries, were common knowledge among women, handed down from grandmother to granddaughter.

The Truth Emerges

It is clear from our study up to now that women were the initial inheritors of spiritual power and authority. The female has a neurophysiological predisposition for mystical and shamanic experiences. The Goddess is complete unto herself. From out of Her being She creates all things.

With the advent of patriarchy the human brain or left brain became dominant and disconnected from the other aspects of the mind. Women, who enjoyed a natural tendency to experience mystic and spiritual pleasure without the need of male priests were consequently persecuted, and in most cases burnt. Sexuality, being linked in women directly with spirituality, became sinful and sexual practices came to emphasize the orgasmic fulfilment of the male only. In some cultures female genitals were disfigured so that sexual orgasm was physically impossible.

The serpent, who was once favoured as a Son of God and Light bearer in all major cultures became evil. The synthesising power of the feminine reptilian mind was repressed for, if it were not, the male, left-brain would be exposed as only a part of a greater whole rather than a dominant factor ruling all.

The reflection of this imbalance of mind is visited upon nature as all the problems with the ecology that exist today. The disconnected dominance of the left brain has resulted in our separation from nature and **other forms of Life which we depend upon for our own Life.**

It is time now for the serpent to rise up the world spine, uniting left and right hemispheres of the world brain. The Anima, the feminine within man, is awakening as the logical structures that man has created breakdown. Each man must know the Goddess within and each woman stand strong in her gift of glory.

Throughout this exploration of myth and symbol a common thread has emerged. Venus, the planet and the Goddess, is directly connected to the root mystery which we need to embody to ensure our continued life on Earth. She weaves a spell, binding

Son and Serpent together around the mother who encompasses all. She tames the collective serpent, gives it wings and teaches it to fly. Her powers of synthesis unite our feelings and intellect; Her Love unites us, male and female, Heaven and Earth.

3

The Alchemical Goddess

> *"For they shall find pleasure, but not her alone;*
> *Seven are her sisters and the least of them is more*
> *beautiful than pleasure".*
>
> *The Prophet*
> *Kahlil Gibran*

The inner male and female aspects of our beings are now being united in fluid synthesis, aided by the stimulation of the Goddess. This merging of gender within the individual is an *Alchemical* change. The main principle of Alchemy is to separate and purify the prime elements of Life, the body (salt), soul (sulphur) and spirit (mercury) and then re-combine them, forming a purer, more integrated whole. Internal alchemy works within, to define and purify the various aspects of our lives; mind, body and emotions and allow the formation of a more wholistic person. Inner male and female aspects are an essential component to this process of purification and integration. In our present phase of evolution, the influence upon the mind of the Goddess Venus/Aphrodite and Her associated archetypes is the paramount influence which can rebalance the distraught human psyche.

The inner synthesis of feelings and intellect, Love and will, is the principle gift that the Goddess brings to us all. Through this inner alignment of the personality the Goddess provides care, sensitivity and intuition in men and allows women to extend their levels of assertiveness, mental clarity and physical strength. All of us are essentially androgynous beings; hermaphrodites in the true sense of the meaning.

We are male as Hermes - teacher of Wisdom, of art and of science. We struggle with our hearts and conquer Typhon, the Dragon of our animal nature.

As Aphrodite we are - female, beauty, Love and passion. We love with our hearts and open ourselves to the wonder of surrender.

Together we are Hermaphroditus - Love and Wisdom united. Both left and right hemispheres of the brain are synthesized together and we express balanced thought and feeling with transcendent wisdom.

The symbolic union of Hermes and Aphrodite in Hermaphroditus can be seen as a blending of both sexual qualities in one being - Androgyny. The androgynous state is a psychic blending of gender qualities within the one individual and a conscious acceptance of the interplay of the masculine and feminine aspects within the individual psyche. Essentially, the New Age of humanity calls for a reappraisal of androgyny, a coming-to-meet between inner male and female values within each individual. Wholeness is achieved through this synthesis. Our separation from nature with all its incidental hazards to life can only be reconciled by inner tranformation of the human psyche so that people (essentially men) can both feel and think without separating the processes from each other.

The missing element for this internal sexual union in our current unbalanced group of societies is Aphrodite, the alchemical Goddess of Love. Within the global dominance of the patriarchy and the physical/technological exploitation and rape of nature for her resources, the feminine qualities of care and sensitivity to feeling have been forgotten. Aphrodite, who was later known as Venus, embodies the pure potencies of the feminine, and, in a world so polarised toward the macho side of male, Her magical Love, sensuality and beauty are irresistible. She is the whore who tempts the priest, the lover of statesmen who brings down governments, the Muse of the poet and artist who leads him crazed and wanting in artistic fervour for days without sleeping. She involves Her subjects in a total magnetic vortex of sensual pursuit for the most sublime pleasure. Kingdoms are surrendered at the gates of Her temple for just one view of Her unveiled self.

Her effect on men is one that causes the most eloquent to stumble for words. She draws them like moths to the flame of Her electrifying, sensate focusing, Love, passion and eroticism. The mass cultivation of Aphrodite in women will lead to the total disempowerment of the macho-war machine. Though She is despised in Judeo-Christian, Moslem and other patriarchal cultures, She is also feared, for every tower of false authority will lean towards Her presence. She is the whore of Babylon. Complete and potent, She will devour the beast that spawns the destruction of nature, Her kingdom.

Though Aphrodite is painted red or scarlet as the whore it is really the judgement of men that makes Her evil. They fear their own weakness and despise Her power which brings even Ares, God of war, quivering to Her bed. Though they may rape Her sisters She holds them spell-bound and powerless. Most hated and loved is She.

In men Aphrodite awakens the most thought-crippling, reason-destroying power ever known; pure romantic Love and sensuality. It is said that behind every strong man is an even stronger woman. Aphrodite represents the power of women over men. She can change the course of history through a short whisper into Her captive husband or lover's ear.

In the case of the more sensitive, artistic male Her effect is like the sip of golden ambrosia that, once tasted, makes all other drinks bitter. She leads the pilgrim on his journey for the grail, to taste once more Her golden nectar. The mundane events of common life are no longer satisfying to the Love-struck man who is embraced by Her beauty. He will crave the taste of Love forever once its wonder is released into his life.

Psychologically, Aphrodite activates the sub-conscious feminine within men; the Anima of Jung. She is the Love, magnetism and attraction which draws a man deep into his psychic and intuitive nature. The primary focus of the male is objective; physical and mental concern for the world of form exterior to his being. Aphrodite is the mystic muse who surfaces upon his consciousness as a moment of sublime spiritual rapture and then disappears into the deep caverns of his subconscious mind, like a bride waiting for him in Her bed chamber. Once awakened to the mystic urge he will pursue Her, discovering the latent mysteries of his own psychic and spiritual roots. Her power to captivate him eventually leads him to the ultimate surrender and fusion of his male and female identities. In this internal Love-feast his soul is brought to the climax of realization. Ultimately all men must make love to the Goddess within.

Though Aphrodite stimulates the passionate Love and erotic nature, that is not the full extent of Her influence. Love and romantic attraction are merely the surface effects of a much deeper, mystical purpose. Love is the bait; the lure which draws the captive deeper and deeper into the shadow of the of the unconscious. The elements of self that are denied; the dark shadow of the hidden closet are revealed and brought to Light by Her Love. She penetrates to the very roots of being and connects the little self with its spiritual power, the soul. Pure Love is the eventual prize, but first comes purification. The Goddess ensures that the captive does not escape - Her Love holds him until the process is complete. When all fear, self-hate and self-denial are purged from the being, the Goddess has completed Her work.

For many the fear of being exposed outweighs the attraction and the prize. The Goddess is despised for revealing weakness in the arrogantly devout. False authority will always persecute Her, for they cannot approach Her in pleasure, but only in great humility and pain. There is no pity in Her heart. If you cannot meet Her love with a pure heart, you cannot approach Her. She comes and reveals Herself and withdraws again. If you follow you will be drawn to dash yourself upon Her rocks. No vessel is perfect enough to contain Her Love, yet still the siren sings.

In women, Aphrodite releases them from the bonds of self-repression and male oppression. She cultivates and refines the senses, opens up a deep feeling of self-beauty and self-worth, allowing women to feel good about their bodies. The golden nectar of Love/electricity and sensual magnetism then flows into women, opening their hearts for pure transmissions of the Goddess' energy. This flood of golden essence awakens and heightens all the senses, both physical and magical, allowing the woman to become a true priestess, embodying the Goddess in her being.

The plot of the patriarchy to destroy the self-esteem of women; to make her sexual expression evil; to limit her freedom and control her body and powers of sensuality and procreation is thwarted once Aphrodite has entered the woman's body. She becomes ecstatic, passionate, artful and magical and will not be oppressed, nor will her desires be denied. Once woman radiates with the Goddess' power men cannot again contain her; she is like a fire, free and expressive.

Psychologically, Aphrodite brings out the assertive, willful and powerfully passionate nature within women. She empowers the Animus, the male subconscious within, like a warrior collecting men's hearts on the point of Her spear. Yet She is not war-like, but psychically penetrating; unveiling and disarming Her potential opponent and lover ready for consumation. However, it is not power and dominion that is sought, but rather a merging and engulfing of Her partner into the vast sea of Her desire.

The Magical Marriage

The revelation of the subconscious as a lover within leads to inner tantric possibilities unimagined in an exclusively singular gender psychology. The interaction between psycho-sexual centres within the individual consciousness leads to ecstatic release of spiritual creative potencies within the individual. The biological woman is more able to embody this internal androgyny because she operates effectively, utilizing both sides of her brain. The male, as was noted earlier, is more isolated in the left hemisphere, largely due to strong societal and peer pressure which casts men into a macho-male role. The development of feminine qualities in men is a powerful taboo, due to the fear of latent homosexual tendencies within the male psyche. Men are unable to express affection towards other males, favouring the spirit of competition rather than co-operation. Because of this, men are largely unable to express or feel Love deeply, leading to a general lack of fulfilment for women in heterosexual partnership. The externalization of this lack of Love and separation has led to our present environmental crisis and only a deep healing of the male psyche can hope to arrest the destruction of life.

Women have always been the initiators of deep feeling in men. Only by breaking the bonds that keep women disempowered and self-critical can the full power of Love be released within our world. Men have lost true spiritual authority through lack of true Love. Instead, they abuse the Earth like spoilt children exercising a false authority through political power, war and economic greed. The patriarchy which condones the destruction of Life, primarily for individual, economic advantage, has, from the beginning, realized the power of women to expose their lie. For this she was and still is in many countries persecuted and denied her sexual and emotional fulfilment, and forced to be subservient, kept uneducated and confined to a strict, gender-orientated, domestic role in society.

Fortunately for Life on Earth she is awakening, rediscovering her latent power and exposing the male lie. Her power is ecstasy and spiritual fulfilment; the synthesis of the inner male and female essences within her dynamic, radiant being. Yet men too have the Light of the Goddess within, deep beneath the veils of the unconscious. She longs to meet the forceful arrogance that leads Her blindly towards desecration. It is Her spirit, rising like a serpent within him, that brings the ecstatic embrace of inner alchemic union.

> "And God created man in his own image, in the image of God created he him; male and female created he him".
>
> *'Genesis'* 1:27

Clearly, even to the grossly patriarchal translation of the Hebrew scrolls in the time of King James, the feminine essence was implanted into the nature of man. Why else is "male and female created he him", included in the text?

The Empress

The Tarot trump 'The Empress', is a clear archetype of Venus and the powers of the alchemical Goddess. She bears the glyph of the planet Venus upon Her throne which is set amid a luxuriant garden representing nature, Her realm.

Upon Her head She bears twelve stars, once again the symbolic drawing upon the Glory of the 'Mazzaroth' or twelve constellations of the zodiac, as we saw in *'Revelation'* 12:12.

> "And there appeared a great wonder in heaven; a woman clothed in the sun, and the moon beneath Her feet and *upon her head a crown of twelve stars.*"

In this mystic symbolism the Empress represents the union of the twelve mid-brain pairs, the neo-cortex connecting both sides of the brain. 'Clothed in the Sun' refers to the Golden Light or nectar associated with Aphrodite and the experience of sexual and spiritual ecstasy.

The **orb** that **She holds above Her head** is the new Earth, the continuous evolving forces of nature, ruled by Venus, that create the seasons and the courses of planets upon the sea of the deep unconsciousness of space. Like the orb of Venus it moves across her face just as Venus moves across and behind the Sun.

In humanity She represents the unconscious realms of the Anima and Animus, the inner female and male counterparts of the deep psyche.

Tree of Life

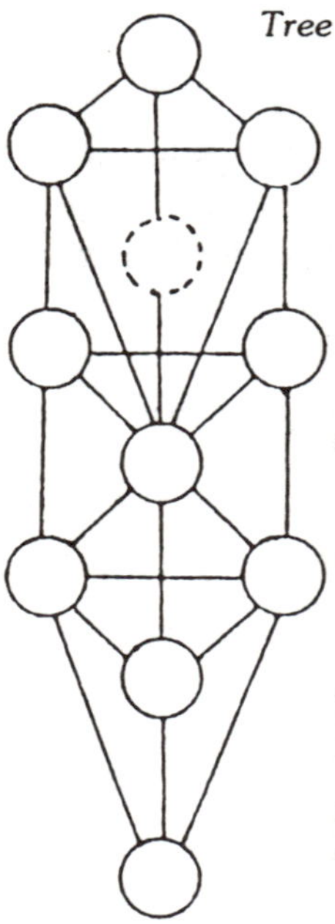

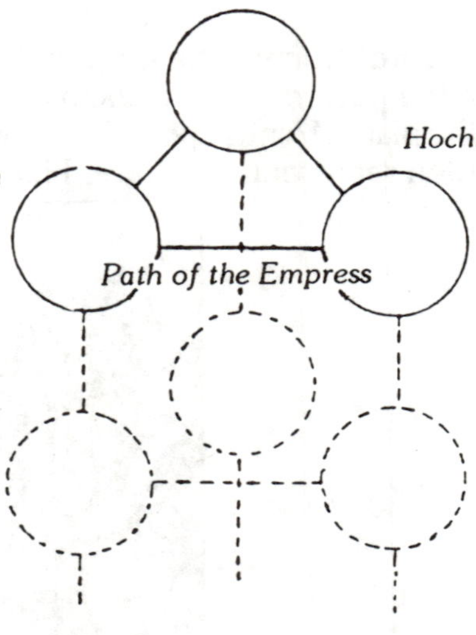

Tarot and the Kaballah

Each of the Tarot cards is attributed to a pathway on the Tree of Life, a mystical symbol representing the creative power of Heaven and Earth and the pathway of revelation within humanity. The Kaballah was pre-evident to the Hebraic scrolls within the cultures of Mesopatamia and Northern Africa as a symbol for cosmic alignment.

The pathway assigned to the Empress connects the upper spheres of the two side pillars of the Tree. (*See diagram, previous page.*)

The whole of the symbolic Image of the Tree of Life contains twenty-two pathways connecting the ten spheres of creation, the Sephira.

In meditation the Tree of Life becomes a body Image of Light corresponding to the physical body in the following way.

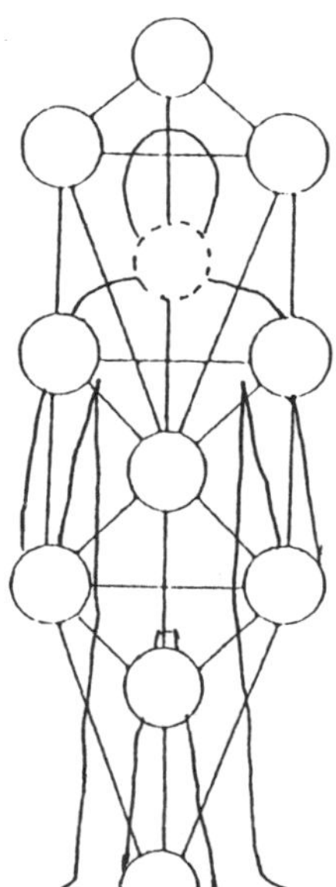

The path of the Empress connects the two sides of the brain represented by the Sephira of Hochma (wisdom) and Binah (understanding). Hochma is the male aspect of the higher mind through which clarity of spiritual impression is registered into cognitive reality. Binah is the dark sea of the unconscious; the feminine right brain which is the blood and form from which idea is realized in materiality. The pathway between them, the Empress, is the magnetic flux that binds Divine Mind and Holy Spirit into creative partnership. The Divine unconscious Anima is likewise brought into relationship with the Animus by the virtue of the Empress' attractive magnetism. Male and female unconscious elements are united by Her power which is the root of all love, affection and desire for union.

Hochma and Binah are the supernal Father and Mother, crowning the masculine pillar of mercy and the feminine pillar of severity. The pathway between these primeval powers represents the deepest union of polar potentialities on the level that they are first manifested from the undifferentiated Light. What Divine Love is this?

While Venus is inscribed upon the throne of the Empress, the planet also holds another place of attribute on the Tree of Life in the Sephira of Netzah lower down in the Tree.

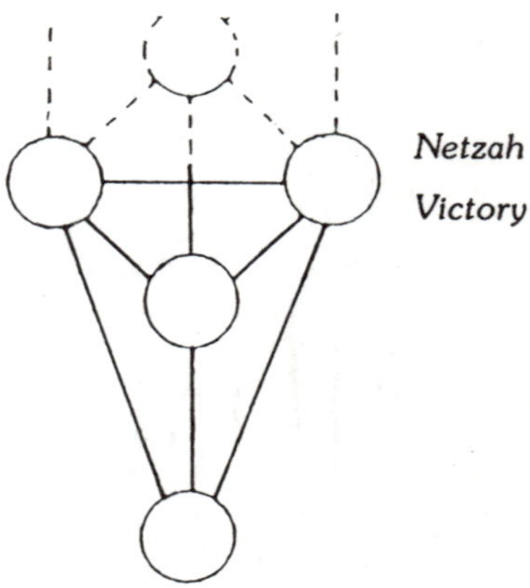

Netzah is in the pillar of mercy and corresponds to the left hip and sacral centre within the physical body. Here the Love and magnetic nature of Venus is cast within the human area of sexuality and generation. The magnetic flux of Venus is now focussed to create evolutionary forms and is polarized onto the active side of the Tree. The power of Venus is here expressed as emotional power, primarily as Love which leads to sexual union. The force which united the supernal Mother and Father in the Path of the Empress now ignites the flame of attraction and desire within their biological counterparts, man and woman.

Netzah and Programme Victory

Netzah is also titled Victory, a name that some relate to as being the victory of morning over darkness, Venus being the harbinger of the new day. However, as a contemporary cipher, 'Victory' has become sacrosanct as 'Programme Victory', the triumph of Light over darkness in our current astrological age. The elements of this higher plan for redeeming the Earth and Her creations are clearly defined within the attributes of Netzah, or Victory. Love as the highest and most powerful emotion, nature as the critical connective web sustaining life, and sexuality as the transformative evolutionary agent, all combine to facilitate passage into a new 'Golden Age'. When male and female are in balance and expressing pure radiant Love, the 'Golden Light' of Aphrodite encompasses their auras. This is the meaning of a 'Golden Age', when male and female are in balance and the world is at peace.

Love is the redeeming agent, capable of transforming the world as no other alchemical agent can. Programme Victory, working through the sphere of Venus, is the paramount source of the Love energy that is now acting to bring peace and co-operation into the world. Victory is a call sign for a process of cyclic renewal that is now taking place; an essential formula for the aversion of global calamity which parallels the extinction of the dinosaurs. Love, as the principle agent for the transformation of humanity, is implementing a union of consciousness and being within the races of homo-industrialist, not just with each other but with the planet itself. Humanity is being drawn by necessity into the realization of the interconnectedness of all Life, primarily as an intellectual understanding, but, as the cry of Victory grows louder, as a total merging of mind, body and emotion; a mass transcendental rapture unprecedented in human history.

Nature is the sacrificial fabric which is reflecting the dire need for humans to awaken to Aphrodite's golden glow. As species by species passes away into the eternity of extinction, an inner feeling of loss surfaces within the heart of humanity. The unprecedented destruction of the Life-sustaining ecosystem of the planet Earth by humans, to the point of being a radical threat to the continuation of Life itself, has broken through the perceived separation that for centuries has veiled human perception in a cloud of selfish unreality. It is the threat to their own lives that is forcing people to awaken as supplies of water, food and air continue to become polluted and unpalatable. We cannot burn down the kitchen and expect to eat in the morning.

The perception of humanity is now focussed more than ever upon understanding the balance of nature and helping to repair and sustain her life-giving properties.

Sexuality is the alchemical agent within humanity which is triggering the conscious embrace of the changing evolutionary tide. The real issues of sexual equality and freedom from sexual repression are the basis for establishing the nurturing conditions through which the golden glow of Love can grow. As more and more people conceive within the encompassing orb of pure Love, the genetic fabric is energised with a renewed evolutionary vitality. Humans, enriched by the capacity to merge and unite totally with each other will also embrace totally the understanding of their planet as a conscious, vastly intelligent entity. The children that are born through the process of making true cosmic Love will be endowed with a capacity to be totally grounded as fourth dimensional souls and spirits within physical vehicles. The increased awareness that these new individuals channelled into the group mind will make all the difference as the ecology crisis deepens. The fabric of their blood and beings is transformed by their parent's capacity to truly align with the vibrations of Victory during conception. These children of the 'Golden Age' will possess sensitivity and a capacity to access information directly from the fourth dimension that will advance the kindergarten of empirical science into the Light technology of the New Aeon. The separation between spirit and matter will be no more.

The Source of Victory

Victory is a direct result of the solar system and the Earth entering into a new band of Galactic radiation as part of the 26,000 year cycle of the progression of the equinoxes. The bio-magnetic nature of humanity, the electro-magnetic nature of the Earth and the solar system, and the quantum mechanical processes of varying star fields all couple together to form a complex matrix of intelligently designed interactions. For instance, the recent discovery of new particle emissions from the neutron star Hercules X1 which have left physicists baffled due to their strange interactive behaviour when entering the upper atmosphere is one example of the constantly changing energies which the Earth is being subjected to. The evolutionary process is cyclic and new revelations in spiritual science will draw clear pictures of how cycles of influence change the fabric of Life on the Planet.

Further evidence on the unprecedented nature of the changes we are going through are evident in the Mayan calendar system which has recently been brought to light by Jose Arguelles in his book *'The Mayan Factor'*. Their pre-computer age astronomers developed the most accurate systems of astronomy and mathematics known to any early civilization. In particular, their reference to a 5,125 year cycle which culminates in the year 2012 AD is relevant to the phase of evolution and crisis which could be interpreted as the transition into programme Victory. The Mayans, according to Arguelles, saw this 5,125 year cycle as a 'Galactic synchronization beam'. The human DNA has been slightly out of phase with the development of Life throughout the Galaxy and during the last 25 years of the beam, the collective wave-form of humanity is to be re-aligned with the 'Hunab Ku' or central Galactic Intelligence.

4

The Return of the Dove

Within the circles of the heart
The white bird flies
From life to life she rides the winds
She's never born and never dies
Sweet soul, your journey's just begun
Sweet lover of the Light your time has come

Daniel Mann

The mystical symbol of the Dove, representing the Holy Spirit, is also directly connected to Venus and her mysteries. The Dove is sacred to the Goddess and a symbol of Her peace giving qualities of Love.

Yet the Dove has become more than an abstract symbol in recent times. The white bird has risen to become the Holy Spirit returning to the consciousness of humanity. From various unconnected sources around the planet sensitives are recording a message of hope and celebration in preparation for 'the return of the Dove'.

> "As the chelas in the physical appearance world have joined their energies and love, in gratitude, through the assistance of their own Holy Christ Selves, the Light from their hearts has formed an upreaching triangle which entered the Temple itself, there to be welcomed into the descending triangle of the Assembly to form a blazing and perfect star honouring Our so-loved Regent Sanat Kumara, the Saviour of the Earth and His lady, the beautiful Venus.

One also sees at this time in the central, uppermost part of the Dome of the Triangle, like unto a vaulted ceiling, a glorious Dove of effulgent White Light, giving forth a radiation of the softest, most delicate shade of pink, the Insignia of the Holy Spirit.

The Cosmic Moment has come and the Divine Edict fulfilled."

'The Gnosis and the Law'
Tellis S Papastavro. Page 114

Regent of the Morning Star

The vision of the Temple of the Holy Spirit represents the personification of the Venusian energy through the dual-Divine persons of Sanat Kumara and Lady Venus. Sanat Kumara has had an extensive relationship with the Earth and humanity according to the teachings in *'The Gnosis and the Law'*. His image represents the anthropomorphism of aeons of magnetic and biological Light vibrations working within the fabric of Life to perfect the seeds of creation. The influence of Venus and Her planetary Regent is the principle factor in shaping life on Earth.

Each of the planets has a Regent, a divine entity who personifies the energy field of the planetary body. The Regent of Venus is variously known as 'Shukra' in Vedic scriptures and 'Sanat Kumara' in the trans-Himalayan texts of *'The Secret Doctrine'* and the writings of Alice Bailey. Shukra was said to have incarnated upon the Earth as 'Ushanas' to give the Earth perfect Laws. Another name given to the Regent of Venus was 'Orai'. The Regent of Venus is in direct astral and psychic communication with the Earth and its spiritual guides, furnishing the spiritual faculties available to the human race.

In particular, Venus is the source of the Holy Spirit or Ghost. As the higher aspect of the divine feminine, Venus represents wisdom or 'Sophia', the spark of pure mind. The lower aspect of the divine feminine is the Moon, the instinctive, form-supporting and regenerative principle that guides the personality in a downward spiral. Sophia, on the other hand, enhances the vision, spiritual growth and understanding of the soul, allowing the higher receptive centres to become active. Blavatsky sees clearly the connection between Spirit and Venus in Her *'Secret Doctrine'*.

> "... Venus (the Holy Spirit), Sophia the Spirit of Wisdom, Love and Truth, and Lucifer, as Christ, the 'Bright and Morning Star".
>
> *'Secret Doctrine'*
> Volume IV Page 110

A further connection is noted elsewhere in this monumental text:

> "In the great Valentinian Gospel 'Pistis Sophia' it is taught that the three powers emanating from the Holy names of the three Triple Powers, that of Sophia (the Holy Ghost according to these Gnostics - the most cultured of all), resides in the Planet Venus or Lucifer."
>
> *'Secret Doctrine'*
> Volume IV Page 81

The Holy Spirit, which is Sophia, is distributed within our solar system by Venus. For this reason Venus has a direct connection to the Earth and human enlightenment. The sharing of spiritual light by Venus is well documented:

> "Light comes through Shukra (Venus) who receives a triple supply, and gives one third of it to Earth".
>
> *'Secret Doctrine'*
> Volume III Page 42

The 'triple supply' refers to the pure emanation of the Light, Word and Spirit which flows unimpeded to Venus from the Supernal. The Earth receives only a third, the Holy Spirit, which is our apportioned share of the emanated glory. Venus is our elder sister upon whom we rely for access to the cosmic stream. For this She is regarded highly in the *'Secret Doctrine'* :

> "Therefore the two are called 'twin sisters', but the spirit of the Earth is subservient to the 'Lord' of Shukra."

and:

> "Venus is the most occult, powerful and mysterious of all the planets; the one whose influence upon and relation to the Earth is most prominent".
>
> *'Secret Doctrine'*
> Volume III Page 42

The cosmic, revelationary influence of our closest neighbour is noted well in the world's mystery traditions. Venus is said to be the 'soul' of the Earth, our planetary 'higher self'. It is this capacity of the solar system to specialize its intelligent evolutionary functions into specific bodies which is at the root of the mystery of consciousness evolution.

Shekinah - Essence of the Dove

The return of the Dove, the Holy Spirit presence or 'Shekinah' is the function of Venus in our present planetary play. The Shekinah or Holy Spirit is the feminine aspect of divinity; the 'third person' in the trinity of Father, Son and Holy Ghost. With Her attributes of spiritual wisdom (Sophia), heavenly Light and divine indwelling presence, She is recognized as the true creative spark within matter and animator of Life. She is the eternal feminine who brings forth life from out of Her being. Our recent emphasis on God the Father as the creator, robs our lives of Her gifts of spirit and Life. Our prostration before the sterile Father causes us to see creation as the manipulation of elements rather than the spiritual evolution of Life.

The Gnostics identified the Holy Spirit (Shekinah) with Sophia (Wisdom). In the Gnostic Creation myth Sophia descends into the waters below the firmament in the form of a Dove. It is not the male deity who is the creator but the feminine Spirit, Sophia, Shekinah or Elohim. The early translators of the Bible, for reasons we can only assume arise from the plotting intentions of the patriarchy, have crowded out the truth that God is both male and female. We are even led to believe that the Holy Spirit is masculine, as the 'comforter' sent after the Son. But, as S.L. MacGregor Mathers points out in his Introduction to *'The Kabbalah Unveiled'*, Spirit is actually feminine.

> "But the word Ruach, Spirit, is feminine, as appears from the following passage ... 'Acath (feminine, not Achad, masculine) Ruach Elohim Chiim': One is **She** the Spirit of the Elohim of Life".

In Egypt the Dove is given the name 'Ba' meaning a vehicle of Spirit. So deeply entrenched was the symbol of the Dove in the time of the early Christians, that, rather than trying to purge the symbol from the minds of their pagan converts, the Dove was given as an emblem of the Holy Ghost and often appears in religious art. The Dove is sacred to Venus, Cybele, Isis, Astarte, Mylilta and of course Aphrodite. It is the Dove, as the Holy Spirit, which represents the creative activity of Deity and that part of the divine from which creation is formed. The direct experience of God can only be experienced by humanity through the feminine spirit. The priesthood of the patriarchy suppressed all mysteries which pertained to the feminine, lest the common folk realize that direct spiritual contact is possible, and that the mediation of priests is unnecessary. The subsequent denial of our only vehicle for experiencing Divinity has left the heart of our souls empty and wanting and led to barrenness of spirit in humanity and our separation from the mysteries of Life and Nature.

A New Age of the Holy Spirit

The modern resurgence of femininity and the mysteries of women are the first signs of a new age of humanity. As Dr. Hurtak so aptly writes:

> "At this time the righteous will understand that just as the Old Testament was the Age of the Father, the New Testament was the Age of the Son, this New Age is the Age of the Holy Spirit."

'The Keys of Enoch'
J. J. Hurtak
page 413

As the flood of Spirit enters our lives, direct access to the Goddess within is becoming common place. The awakening and revelation of the Holy Spirit in humanity will lead to the eventual realization that we are one people. The separation of humanity symbolized by the myth of the Tower of Babel will end as our consciousness merges again with the creative spirit of Life and Nature. The common root of life binds us all to the consciousness of the Earth. Her devastation and rape by a humanity cut off from its roots by the cruel axe of patriarchy is the common tear that will unite the people of the Earth. Our future is dependant on the reversal of our primary attitudes toward Life. Instead of seeing ourselves as superior and separate from nature we need to recognize our connectedness to all forms of life and use our intelligence to serve the wonders of creation rather than to destroy them.

The Pleiades - Constellation of the Dove

Many of the mythic fragments that have survived from ancient times have a direct link to astronomical and astrological cycles. Knowledge that would otherwise be lost is preserved in myth and legend. Even the great monuments of stone erode when subjected to the elements of time. The Great Pyramid was once covered by a limestone sheath bearing symbols. The stone monuments of Megalithic Britain have been cast down by farmers. Yet, as the stone calendars of Central and South America bear witness, the science of astronomy was well advanced in ancient times.

The Pleiades, seven stars adjacent the constellation Taurus, were represented as a Dove in Egyptian Art:

> "Now, in Egyptian sculpture, or painting, the Pleiades are sometimes represented by the figure of a dove with wings outspread over the bull's saddle. The dove as we remember, is the bird sacred to Venus, and as the Pleiades are part of the constellation of Taurus and, as we shall see, more Taurean in nature, if possible, than Taurus itself, the dove becomes a specially appropriate symbol for this little star-group".

'The Zodiac: A Life Epitome'
Walter H Sampson
Page 24

The mysterious connection between Venus and the Pleiades is well noted in the above passage, for the Pleiades are the source of the spiritual energy which Venus represents in our solar system. The cosmic connection between Venus and other stars in our galaxy was known to early wisdom religions:

> "It is significant that in the pagan mysteries the dove of Venus was crucified upon the four spokes of a great wheel, thus foreshadowing the mystery of the crucified Lord of Love".

'The Secret Teachings of All Ages'.
Manly P. Hall, page XC

The 'great wheel' is of course the zodiac and its 'four spokes' are fixed signs: Leo, Aquarius, Scorpio and Taurus. In many of legends and myths of primitive humanity the heros and heroines are often set in the stars when they die, thus forming the mythological nature of the various constellations. The four beasts of the fixed cross are given in the vision of the Cherub of *'Ezekiel'*, representing the four rivers of Life flowing from the Creator:

> "As for the likeness of their faces, they four had the face of a man, and the face of a lion, on the right side: and they four had the face of an ox on the left side; they four also had the face of an eagle".
>
> *'Ezekiel'* 1:10

The lion represents Leo, the man's face Aquarius, the eagle Scorpio and Taurus is the ox or bull. Taurus is always the sign of creative force, and connected with Christ, who was proclaimed the Light of the World. In Egypt the sign of the bull was known as "the interpreter of the divine voice", and, as the Christians phrase it "the word made flesh". Jesus, we have seen, is directly connected to the mysteries of Venus, his doctrine of Love being the principle energy of the Morning Star. It is interesting to note how the mission of his life came upon him in the shape of a dove. John the baptist begins speaking of Jesus who is yet to come:

> "I indeed have baptized you with water; but he shall baptize you with the Holy Ghost.
>
> And it came to pass in those days, that Jesus came from Nazareth to Galilee, and was baptized by John of Jordan.
>
> And straight away coming up out of the water, he saw the heavens opened, and the Spirit like a dove descending upon him:"
>
> *'Mark'* 1: 8-10

The path of Jesus is from the outset marked by the dove and the Holy Spirit. The gifts of healing and wisdom that Jesus exhibited in his life were only made possible because of the Holy Spirit, the divine intermediary and only way in which humans can know God. The illumination by which Jesus became "the Light and the way" is the central theme of the constellation Taurus and the stars of the Pleiades.

The brightest star in the constellation of the Dove is Alcyone, meaning, "the one eye of the all seeing One". Alcyone is believed to be the metaphysical centre of our Galaxy. The Arabs called this star Al Wasat, the central one, and the Hindus refer to it as Amba, the mother, whose milk flows out to become the milky way. In Hebrew tradition the Pleiades are known as Kimah and quoted in *'Job'*:

The Pleiades

> "Canst thou bind the sweet influence of Kimah or loose the bands of Kesil (Orion)".
>
> *'Job'* 38:31

The 'sweet influence' is the feminine creative Spirit which weaves Light into form. The seven stars of the Pleiades are called the seven wives of the Rishis, the original male power symbolized by the seven stars of the Great Bear in Hindu scriptures. The influence of Ursa Major, the Great Bear, is the power which has brought about the warlike domination of male energy upon the Earth. This can only be balanced by the 'seven twinkling sisters' of the Pleiades. The seven sisters, as the Pleiades were also known by Australian Aborigines, nurture and enhance the powers of creation while the Rishis of the Great Bear are the destroyers of form. Our attunement to the feminine and the givers of Life, the Pleiades, is the transition necessary for us to enter the Golden Age of Light.

The following extract may shock some of you and you may not believe these words to be true. We are about to enter an age of Light and of the Holy Spirit. The following article appeared in the magazine *'Phoenix Rising'*.

The Photon Belt New Age

"The New Age has arrived when 'Heaven rests on Earth'. Is this the New Heaven and the New Earth?

It is anticipated that the Earth's rotation may diminish a little. Because of the now reduced solar radiation, the temperature will drop and the ice caps will start to extend to about latitude 40 in both hemispheres. History books record that at least five ice ages have been recorded, about 10,000 years apart, and that they last about 2,000 years. It may be noted here that all world communication centres, fixed satellites, American bases and experimental sites are all within the 'safe' zone ... design or accident???

What will happen to you now that the Earth is posed to enter The Photon Belt, predicted for around the year 2,000???

There are three types of 'people' in our cosmos. There are the Corporeal, like us, solid, humanoid. There are the Atmospherean, also solid up to a point, but with a different molecular structure. The electrons have moved out a little and emit Light. (Jesus was born an Atmospherean, not corporeal like us, which is why He shone with Light). Then there are the Ethereans, who have no mass at all.

Upon entering the Photon Belt, a normal healthy person is expected to feel a jolt similar to putting a finger into a live light socket, and the transformation from a Corporeal person into an Atmospherean person will be complete. To quote from the Bible: "And ye shall be changed to immortality without the separation of death, in the twinkling of an eye".

Theologians have written at great depth on Bible characters, and assert that they lived in this period of Light, when the sky and atmosphere were different, and apparently it never rained. There was heavy moisture and the streams ran, but it did not rain. Nostradamus, in his quatrain about the end of the world as we know it, said: "And it will rain no more".

In Australian Mythology, it is said "men were different to that which they are now", "we had a bridge to the stars". In all their stories, it is said that if they had a falling out with the witch doctor or chief, they "fled to the skies". So did the Greeks. It seems that space travel is simple within the Photon Belt.

1962 was the year the Earth came within the outer influence of the Photon Belt. It was a year of great UFO activity. Did the Earth come within range of some space travellers within the Photon Belt? As it hurtles closer and closer, will more UFO's watch over us before our rebirth into the years of Light? It would seem from intelligent research that this is happening.

Erich Von Daniken, when visiting South America, discovered a tribe with an object given to them by 'Sky People' thousands of years ago. They were told to keep it clean and when it begins to hum like a swarm of bees, "We will return". This object has begun to hum softly. Is this a cosmic alarm clock sounding the alert for the Coming of the Light?!

It would seem that some civilizations live permanently within the Photon Belt. When this planet leaves the period of Light and returns to the 10,000 years of darkness, do these visitors return to their own planets in the Light and await our eventual re-entry? It seems likely! The Mayans departed hurriedly with the message of their return, which scholars say is now imminent. Rock carvings around the world show drawings similar to the System of Alcione.

Ball lightning is a new phenomena about which very little is known. Is it perhaps outlying pockets of the Photon Belt?

To all UFO researchers, unidentified UFO craft always show an interest in nuclear installations. What will happen to reactors within the Photon Belt. Photon Energy will be the energy of the future. Many UFO reports seem to indicate photon rockets on experimental wingless craft. People reporting such craft say "the headlights were in the back". These reports indicate slow, cumbersome craft, but within the Photon Belt, many nations could have a stockpile of "wingless craft" ready for immediate space travel when the time comes.

The radiation of the sun will be modified by the Photon Energy. Is this why scientists pay only lip service to solar energy? Within 18 years, solar power will become redundant. What does it mean to be an heir to the New Age, or the Photon

Age? If the concept of the Photon Belt can be accepted, then the need to raise our consciousness, to be spiritually aware, as well as to make ourselves physically, emotionally and mentally more healthy, takes new meaning.

The answer to peoples searching is sometimes SO SIMPLE that they look past it, ignore it, and try to find a much more complicated answer, when all the time the answer is ALWAYS there ... their passport out of the vortex into the future. The key to the jigsaw puzzle is a small atom, and a small group of electrons which orbit around it. EVERYTHING that we can see or touch is made up of atoms. For reasons our scientists have not yet discovered, the electrons around the atom move out a little and emit a photon, a PARTICLE OF LIGHT!

This takes us into the world of anti-matter. In the near future, physical life is going to depend on the behaviour of these photons. So let's have a look at this phenomenon. English Physicist Paul Adrian Dirac, said that for each type of particle, and anti-particle will exist. In 1932 Carl David Anderson discovered the anti-electron, and called it a POSITRON. In 1956 the Anti-proton and Anti-neutron were discovered. When an Anti-particle is formed, it comes into existence in a universe of ordinary particles, and it is only a matter of time, a millionth of a second, before it meets and collides with an electron.

The charges cancel. The total mass of the pair is converted into energy which will partially be in the form of photons. This offers a new and unprecedented powerful source of energy. THE PHOTON is about to become the way of life in the very near future ... life or death!

The second part of the jigsaw puzzle is The Pleiades (the Seven Sisters), an estimated 400 light years from here. This group of stars is at the basis of mythology in all countries, the Greek Gods, Australian Dreamtime, Chinese Mythology, to name just a few.

Paul Otto Hesse also made a study of this system and discovered at absolute right angles to the movement of the orbiting suns - a Photon Belt. Our scientists have not been able to reproduce this phenomena (emission of photons from an atom without the resulting nuclear fission), so they refuse to admit publicly the knowledge that this belt of Photons exists!

The Photon Belt was discovered in 1961 by means of satellite borne instruments and our approach to the Belt is monitored by orbiting satellites. It takes our sun 24,000 years to complete one grand orbit. The American Indians have always believed that time is a circle, which always comes back to the beginning, and they call this 'One Day'. The Bible states: 'On the first day God made ... On the second day, God made ...' This now begins to made sense. With each 'day' being 24,000 years, the creation of the world becomes quite logical.

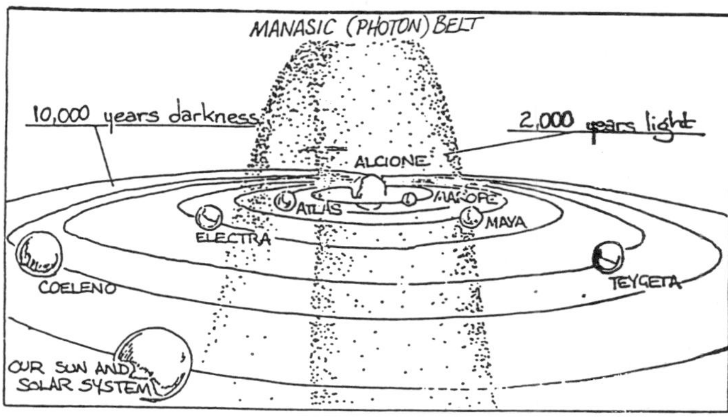

The Pleiades and system of suns of which ours is a part

When the sun has just left the Photon Belt (as seen in the diagram above), it takes 10,000 years to reach it again. These are called the years of darkness, a day and a night, as known to us. Then there are 2,000 years within the Photon Belt, and so on.

OUR SUN AND SOLAR SYSTEM ARE NOW POISED TO ENTER THIS PHOTON BELT! IT IS INEVITABLE!

Sometime between now and the end of this century, this planet will come out of the years of darkness into the Light Years, or 'the Photon Era'. The full circle is nearly completed, as we move back to The Beginning. It is described in detail in the Bible, by all books of mythology, by Nostradamus, and by some scientists.
According to Paul Otto Hesse, if the Earth enters the Photon Belt before our sun does, the sky will appear to be on fire, but

> be assured, this is cold light, like auroras, so there will be no heat. If the sun enters the Photon Belt before the Earth (and the Bible claims support of this) there will be immediate darkness, which, computed at the Earth's present speed through space, will last 110 hours.
>
> The interaction between solar radiation and the Photon Belt will make the sky appear as if it is full of falling stars. The Bible states 'All the stars will fall from Heaven and the sky will be no more'!
>
> As the Earth enters the Photon Belt (as explained earlier) all atoms will become excitated (sic) and their electrons will move out a little to emit those particles of Light. All things will change. All things will become slightly luminescent. There will be constant Light. There can be no darkness, not within the deepest cave, not within the human body. We will become Atmospherean. We will exist in a new vibration. And so tomorrow ..."

The coming Photon Belt is akin to the activity of the Holy Spirit being revealed in the heart of every atom as Light. It appears that the 'Return of the Dove' points to a direct physical occurrence as well as a spiritual experience. Further evidence of imminent changes about to occur in the energetic nature of our solar system can be found in *'The Keys of Enoch'*:

> "There is presently occurring a space-time overlap with 'the Higher Evolution' as the Earth's solar system enters an electromagnetic null zone, a vacuum area in space which will change the magnetic forces of creation..."
>
> Page 169
>
> "Upon entering the lip of the electromagnetic null zone, the whole solar system will be picked up and re-seeded in a different vibratory dimension of the Galaxy ..."
>
> *'The Keys of Enoch'*
> J. J. Hurtak, Page 124

There is no way to describe the changes that are likely to occur as a result of these phenomena. The Mayans recognized the limitations of our current cycle and their calendar indicates its final culmination. Jose Arguelles calls this process Galactic Synchronization:

"To speak of Galactic synchronization, synchronization with the beyond, is to surpass all fantasy and all of our wildest dreams. ET's, UFO's, the 'space brothers' - these are not alien entities, but emanations of **being** itself. And being is, in its essence, light, radiant energy. From light we come and to light we shall return."

'The Mayan Factor',
Jose Arguelles,
Page 146

5

Cycle of the Morning Star

"Earendil, the Evening Star, most beloved of the Elves, shone clear above. So bright was it that the figure of the Elven-lady cast a dim shadow on the ground. Its ray glanced upon a ring about her finger; it glittered like polished gold overlaid with silver light, and a white stone in it twinkled as if the Even-star had come down to rest upon her hand."

Frodo discovering Nenya, the Ring of Adamant, upon Galadriel's finger.

Fellowship of the Ring
J. R. R. Tolkien
Page 346.

One of the most significant factors linking Venus and the Earth is the indisputable harmony of their orbital cycles. In particular, the alignment of Venus, Earth and the Sun in the progressive cycle of inferior and superior conjunctions demonstrates the close affinity of the two planets. When viewed from the Earth, Venus will pass between the Earth and the Sun approximately every 584 days. The time between these inferior conjunctions and the passage of Venus behind the Sun, a superior conjunction, is half of this or 292 days. With the Earth completing one circuit of the Sun every 365.25 days, it is a mathematical inevitability that Venus, Earth and the Sun will return to the same relative positions almost exactly every eight years.

If the relative positions of the Sun/Venus conjunctions are plotted out over eight years upon the circle of the zodiac it will be seen that the figure forms a pentagram!

During the eight years which began with the Venus/Sun conjunction at fifteen degrees Aries on the 5th April 1989 and culminated in the conjunction on the 2nd April 1997 at thirteen degrees Aries, five superior and five inferior conjunctions took place, ten in all.

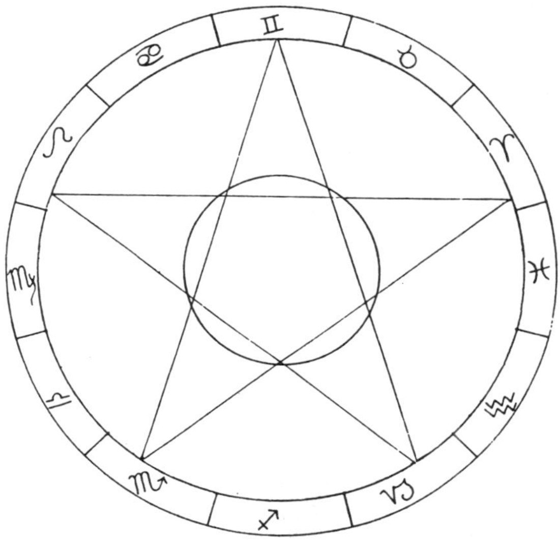

Ten sun/Venus conjunctions plotted on the zodiac

The significance of this pentagram of cosmic proportions is immense. The pentagram has long been the symbol of human perfection, the victory of intelligence over the four elements. Magicians use the inscription of the star as a means of protection and to banish and invoke elemental forces. Its five points represent the five senses, the five digits on the human hand, the arms, legs and head of our body and Earth, Water, Fire and Air presided over by the ineffable Spirit or Ether. That Venus should create this orbital bond with the Earth is a demonstration of Her mystic connection as soul protector of our brilliant blue planet and Her evolving life forms. The nearly perfect geometric harmony in the orbits of Earth and Venus is like a cosmic dance, a partnership of Love and motion that, like a magical rite, seals and protects the order and purpose of our solar system from dissonance and corruption.

The Dual Nature of Venus

Venus is both Evening and Morning Stars and passes before and behind the Sun. The orbital pentagram just described has ten conjunctions, two at each of its five points. One is an inferior conjunction, when Venus is between Earth and Sun and the other is a superior conjunction when Venus passes behind the Sun. Thus we have, in fact, two pentagrams - one Light and one Dark, representing, as in the Tao, complimentary passive and active polarities. As Venus is brightest and closest to the Earth at the time of inferior conjunction, I would consider these to be the Light or Yang aspect and the superior conjunction when Venus is farthest away the dark or Yin polarity.

Symbolically, this relationship between Light and Dark can be represented by the upright and inverted pentagrams.

There are other factors which lend power to the dual expression of Venus. Her cycle as both evening and morning stars is well noted both astronomically and mythically in such legends as Quezalcoatl, who is linked to Venus as both Tezcatlipoca, Lord of Darkness and Quetzalcoatl, Lord of the Dawn.

The inescapable connection between Venus in its dual aspects and our evolutionary dilemma is clear. The human condition is governed by the quest to resolve opposing forces. Within our own beings we face the enduring struggle between soul and personality or as the Kabalists would have it, 'Neshama', pure soul, and 'Nephesh', the animating vital function. One aspect of our beings faces ever upward, striving for Light, Love, Truth and Understanding while the other part strives in ignorance, lusting after material things, wealth, power and greed.

Esoterically considered, Venus is the sidereal symbol for the fifth principle, Manas (mind), which is dual. The vibrational plane of the mind (Manas) is divided into two parts, Kama Manas, the animal mind, spiritually paralyzed by vice, and Buddhi Manas, spiritual self-consciousness. In the evolved being these two aspects are joined by a bridge of Light, the Antahkarana, so that lower mind, the thinking personality, can blend forces with its Divine Ego to dissolve the bonds of material limitation and containment. The true mystery of Venus is the linking together of the split fifth principle, uniting spiritual and material selves. The symbol of the Light and Dark pentagrams fits well this struggle between soul and personality.

Above we have the higher mind reaching upward for Light and below the downward seeking lower mind. Between them is a great abyss, the crossing of which determines the evolutionary direction of the soul. The task is to draw the higher into the lower by activating the force within, the heart, seat of Love - Buddhi. Within the heart of each pentagram, the compliment is hidden. When the inner Life is activated and becomes dominant the inner power attracts lower and higher together as illustrated here:

It is not difficult to see now why Venus and her various deities have been subject to such devotion and also such hatred. We hate our own darkness and we love our own Light. Venus shows us both in her dual Light, morning and evening stars. Yet truly we are **one** being, of **one** substance and through unification by the power of Venusian force, Love or magnetism, we realise the indivisible nature of higher and lower. Lower mind, alone without guidance, bleeds its light dry and is a bane to the soul and its promise. However, when higher and lower meet there is natural accord, purpose and union and lower becomes the willing servant of the whole. There is never any evil as such, only division within the wholeness of being.

Venus and the Fifth Ray

Esoterically, Venus is the planetary body which infuses Earth with the quality of the Fifth Ray of Concrete Science. There are seven great rays which condition consciousness that sweep in and out of manifestation in our solar system. The fifth of these is 'Concrete Science'. The energy of the Fifth Ray seeds the thinking mind with Divine knowledge, as the seeds of light stream from higher to lower minds across the Rainbow Bridge. The motivating energy of this ray causes the mind to visualize and manifest, design and create according to higher inspiration.

It seems a paradox to some that Venus, planet of Love, should be so involved with the transmission of mind and intelligence. Love is only the veil through which the unevolved feel Her calling, inspiring them to fuse lower and higher together through the generation of golden light in the heart. Once the bridge is constructed and the being is fully connected, the true nature of Her force is evident. As the higher self or soul to the Earth, Venus is the 'Revealer of the Way' for Earth-bound intelligence. Once the mystery of Venus and the dual-pentagram are realized, true cooperation between soul and personality can be achieved, both individually and planetarily.

Numerologically the number five is a key to the unification of self. Five stands midway between one and ten, both of which represent unity of being. Five is the separation point and also the place where the separated may be bound together again. Simply seen; five plus five equals ten.

$$5 + 5 = 10 = 1 = \text{unity}$$

The five below, 0, 1, 2, 3, 4 ,are thus united with the five above, 6, 7, 8, 9, 10, once the dual nature of the fifth is resolved.

The Fifth Ray is thus the meeting place between spiritual and material worlds, the influence which unites them together in a creative bond. Further geometrical evidence of the fifth component being the bridge which spans the abyss, passing complex creative signals from above to below is seen in the shape of DNA. When viewed from above the molecule looks like this:

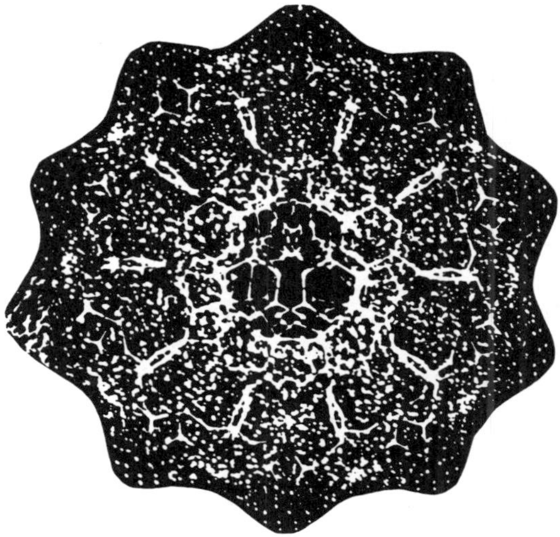

Life itself is thus coded through an interlocking of two pentagrams into a chemical matrix. Intelligence too follows a pentarhythmic harmony in the path of manifestation from macro to micro. The evolutionary impulse is translated from the universe into our solar system by the orbital pentagram set up by Sun, Earth and Venus.

This scientific multi-level analysis that I am suggesting is precisely the quality which manifests through the influence of the Fifth Ray. The channel which connects the physical and spiritual worlds is revealed in precise concepts and scientifically acceptable facts making the flow of the creative impulse completely articulate to the human mind. The Fifth Ray is waxing in strength during our present age, increasing the importance of a conscious link to Venus and Her conditioning intelligence at this time. Humanity is learning the mysteries of the universe, both within and without. As the Fifth Ray stimulates our consciousness we will enter into closer communion with the sources of knowledge in our local universe and come to understand the laws that govern creation. With this knowledge we will be empowered to once again take our place as a member of the evolved galactic community and achieve wonders as yet undreamt of.

The Venusian Hierarchy

There is much mystery surrounding the Fifth Creative Hierarchy - Hierarchy in this context meaning an evolutionary group of souls, bound together by a common cosmic and karmic path. The true soul is forever in communion with the Hierarchy, and we humans perceived this bond consciously only to the extent in which we are able to focus upon the energies of our soul. The Fifth Creative Hierarchy is linked to the planet Venus and the constellation Capricorn. In *'Esoteric Astrology'* Venus is said to be the 'Hierarchical Ruler' of Capricorn. The Fifth Hierarchy is also connected to the symbol of the pentagram. The insights and services to creation rendered by the Fifth Hierarchy educate the empty animal form and encourage growth into rational human beings. The influence of this Fifth organization of souls is to stimulate the development of the Fifth principle, Manas, and initiate consciousness into full connection with the soul and the greater organized intelligence.

Consider this influence to emanate from intellectual and conscious Spiritual Beings through the super-senses. These Spiritual Beings have a special affinity with the magnetic intelligence of Venus and the constellation Capricorn. All planets and stars possess intelligence and are conscious. These vast beings are part of an ordered, intelligent universe, influencing the growth and evolution of lesser beings.

The Fifth Spiritual Hierarchy, the crocodiles or 'Dragons of Wisdom', act as channels for the consciousness of the constellation Capricorn, which is connected with the birth of the spiritual microcosm and also the passing away of the physical. Capricorn guards the secret of the soul and its entrance into conscious life. Venus rules over the Fifth Hierarchy in Capricorn because She is the soul for our planet and it is these combined influences which enable the soul to fully manifest within the body.

The Time of Mystery Unveiled

Evolution of consciousness upon the Earth is not a happen-chance series of accidental discoveries, but rather a carefully staged development linked to the cyclic procession of planetary and solar systemic motion. The keys that unveil the course of evolution are gained by the insights of esoteric astrology, which interprets the symbolic progression of planets through the zodiacal regions into precise patterns of cosmic influence.

The mysterious keys to the hidden art of esoteric astrology have been scattered amid the remnant wisdom cultures of the world. According to the descendants of Quetzalcoatl, Lord of the Dawn and the Morning Star, we are presently in the Fifth solar age, known as 'Navi Olin' which means four movements. Movement (Olin) in this sense is related to the element fire, and describes not only the quality of this age, but the nature in which it will end.

Current indications from both the Mayan calendar and the fulfilment of Hopi and other prophesies point to the culmination of this age by the year 2012. The fiery end of this age can be interpreted in two ways; by either the increase of destructive powers from within the Earth - such as earthquake, volcanic eruption and, in humans - war and disease or, Olin can manifest as 'Olin Yoliztli', the movement of the sacred heart.

The end of this fifth solar age is also tied in with the culmination of the Christian *'Revelation'*. When John the Baptist spoke of Jesus he said that Jesus would baptize not with water, but with Fire and the Holy Spirit. The Christ ministry with its emphasis on Love demonstrates the highest use of that fire, not as a means of destruction, but as a fire within the heart.

In the final stages of this age we have an opportunity to choose the way in which the world makes its transition from one Age to the next. We **can** invoke the energies of the Sacred Heart, made manifest through Venus. Through Her we can call for the return of Her promise to the world; masters such as Christ and Quetzalcoatl who are prophesied to return now. The cultivation of the heart as a conduit for the flames of the soul, the true solar fire of being, is the critical redemptive factor; the force which can transmute the elements of destruction into a unified human spirit. Venus in her capacity to engender Love into the human heart is the key influence at this time.

The culmination of the Fifth Age will also bring the Fifth Kingdom, the spiritual hierarchy, into contact with the Fourth Kingdom - humanity. As souls awaken and bring forth the truth of God realization from across the Rainbow Bridge, the full cooperation of the spiritual organization of souls will unite in the perfection of the creative plan. Humanity will become self-realized, which is the goal of self-consciousness, and embrace the spiritual directives that are etched deeply in each soul as individual purpose connects with organized, united effort.

Igniting the Venusian Flames

Many individuals are proceeding to anchor the Light of true Self into their personalities, forging a spiritual connection which is assisting in the progressive en-Lighten-ment of global humanity. There is, however, a moment of cosmic importance emerging for the collective realization of the global heart of humanity. The planetary cycle we have commenced with the entry of Neptune, Uranus and Saturn into the sign of Capricorn is drawing the attention and energies of the Venusian Fifth Hierarchy, which is being empowered, allowing the mysterious force of Capricorn to initiate a new response in evolving humanity.

The cycle of these outer planets provides the necessary opportunity for the 'Dancer from the Deep', Venus, to enter into the human heart, igniting the flame of Love, which draws Heaven and Earth together. The Fifth Hierarchy is brought into manifestation, influencing the evolution of humanity, once the proper alignment of the three synthesising schemes, Saturn, Uranus and Neptune have occurred in the sign Capricorn. Venus, the soul of the Earth, is then able to shed Her true light without obstruction. The key to this mystical blending of planetary and stellar influence is given in the Greek myth surrounding the birth of Aphrodite (Venus). The myth is an astrological code which, when properly interpreted, gives the esoteric astrologer an understanding of when certain cosmic forces are renewing their contact with the Earth. The basic framework of the code is thus revealed:

Gaia (the Earth) called upon her son Cronos (Saturn) to castrate his father, Uranus. When he completed the act he threw the severed genitals into the sea. White foam spread about them as sperm blended with water. Aphrodite arose from the foam, perfectly conceived as a fully grown Goddess.

In esoteric astrology this meeting of the Gods represents a remarkable cycle for planetary renewal. A need within the Earth (Gaia) invoked the desire for Venus to be born into the system. Saturn (on her call) castrated Uranus. In other words the planets Saturn and Uranus were in close conjunction and their opposing energies resulted in the creative spark of potentiality (the Genitals) of Uranus to be brought into the guiding hand of Saturn. Saturn and Uranus have an opposite effect upon the personality. Saturn has a conservative, consolidating effect while Uranus is spontaneous and unconventional, bringing the new and unexpected into manifestation. The organising and disciplin approach of Saturn is thus empowered in this mythic play to conceive with the generative spark of the Father. The previously aloof and unreachable powers of cosmic generation are now structured into the real world, allowing human souls to participate in the act of co-creation. Uranus is the unpredictable electrification of the creative Spirit and ruler of Aquarius and the violet transmutive ray. Saturn's victory over his powerful Father could only be achieved through the aid of a favourable cycle. When Saturn and Uranus meet in Capricorn, Saturn's powers are enhanced due to Saturns rulership over that part of the heavens, both esoterically and from the orthodox point of view.

Saturn then hurls the organ into the sea (ruled by Poseidon or Neptune) which foams bringing forth Aphrodite or Venus. The sea represents the vast unconscious mind. The emergence of Aphrodite from the unconscious is like the release of the feminine from denial within ourselves. The Archetypal Goddess releases the inner female from bondage, resulting in outward acceptance of the feminine and its principles. In March 1989 the first of three exact conjunctions of Saturn and Neptune began. By the end of 1989 Saturn would have symbolically thrown the organ of creation into the sea of Neptune. The birth of Aphrodite (love Goddess) occurred once the essences had blended and foamed. The Uranus and Neptune conjunction in late 1993 will symbolize the blending of sperm with foam, culminating in Aphrodite's birth soon after.

The awakening of Aphrodite from the foam is symbolic of the awakening of planetary Love from the present cycle of planets in Capricorn, culminating on January 13th 1994.

At this time Venus will conjunct Neptune and Uranus in Capricorn and the purpose of Mother Earth (Gaia) will be evident. The need of our planet for Love, Peace and respect for Life will emerge victorious. Saturn (Cronos) will have served his mother well, for Love shall be born as plentiful as foam upon the sea. The Goddess will return to us with Her blessing. The rebirth of the Love aspect will construct the Rainbow Bridge for mass humanity. Personality and soul will connect and the higher mind, Manas, of global humanity will begin to wrest control of the Earth from the animal-man.

The once veiled importance of this Greek myth now stands revealed. The ultimate victory of Aphrodite/Venus, the Goddess of pure Love, is assured by cyclic decree. The paternal hierarchies of the Earth will be brought to their knees in awe before the splendour, beauty and ultimate Love of the Goddess as She animates all that is feminine in woman, the Earth and in nature. Then will the rebuilding occur as one world passes away and a new cycle begins.

Solar Fire

The transition of human consciousness from animal-mind to soul-mind can be demonstrated from another point of view using the Tree of Life and its astrological correspondences. Human consciousness is presently evolving from an astral-emotional fixation into full contact with higher mind and soul. Consider the Tree of Life with the eleven 'Sephiroth', or spheres of emanation, which acts as a map between Heaven and Earth. The astrological attributes and basic keynotes are given in the following diagram:

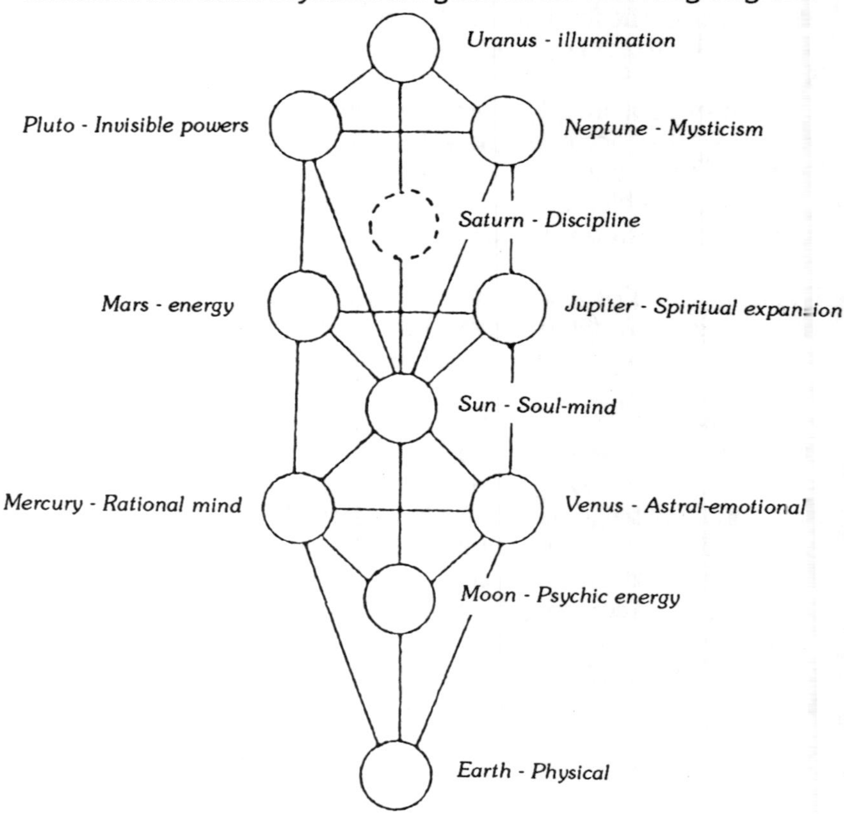

Though all levels are present in the creation of the human form, consciously humanity is generally only operant on the lower level of the Tree.

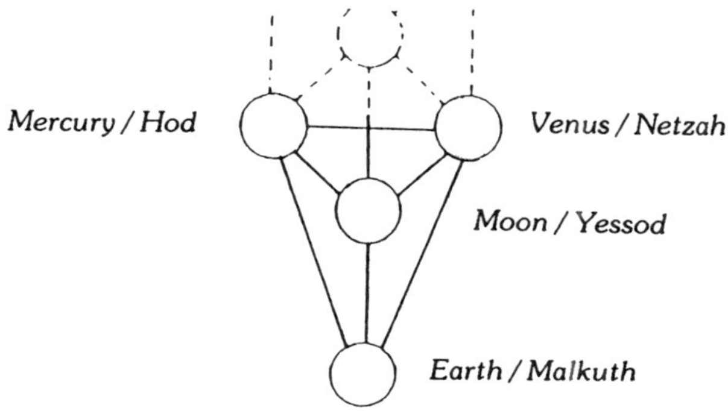

Malkuth/Earth - This is the centre which governs the physical body. Hunger, pain and other bodily signals emanate into consciousness through this Sephira.

Yessod/Moon - Yessod is the realm of instincts, the visual screen that receives images from the imagination and mind. This is the psychic realm which reflects thought forms and empowers their manifestation into form.

Hod/Mercury - Hod is the ever reverberating activity of the mind. Memory and active thought processes of the conscious mind in response to sensory input emanate from this Sephira.

Netzah/Venus - Netzah represents the dream-self, the emotional and astral body at work in translating between conscious and unconscious. In its purity it is the web of consciousness that interconnects between self and all other life forms - the forces of nature.

Normally humans operate in their daily lives in two triads of interplay, either Hod, Yessod, Malkuth while awake or Netzah, Yessod, Malkuth while asleep or regenerating. By stilling the lower Mind, Hod, through meditation or any practices that induce Alpha waves in the brain, the normally unconscious, dream-self, Netzah, is allowed to become active in conditioning consciousness. Thus Netzah and its planetary attribute, Venus, represent a transcendent pathway once they are invoked consciously into the being. The harmonious blending of Self with non-Self experienced through Netzah is an important gateway through which humans can gain access to the higher mind, which shares being with transpersonal spiritual aspects of Self.

If we view the Tree of Life as a pathway of emanation from the creator to the physical, a natural pattern unfolds called the 'Lightning Flash' which represents a gradual, step by step transfer of energy from finer to grosser vibrational levels.

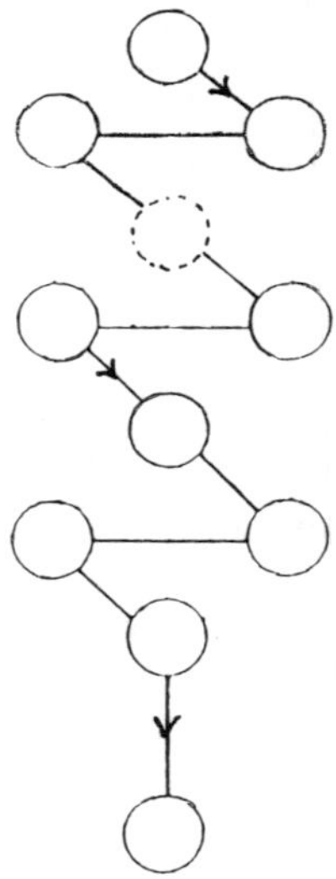

The Lightning Flash

Netzah is thus the first sphere to be emanated into the lower world of substance and action in which humans generally spend their lives. Each sphere is governed by the preceding higher sphere according to this pattern. Thus Netzah causes Hod to become passive, most commonly through sleep. The power of the dream self over the conscious mind is well documented in dream deprivation studies where people who are continuously awakened when their mind goes into dream-state eventually begin dream-hallucinating while awake. Netzah is a higher, finer and more spiritual vibration than Hod. The dream-mind, connected to nature and the emotions, is an elevation in consciousness from the rational, sense driven lower concrete mind of Hod. It is a shame that present day humanity is dominated by the lower mind while awake as this has led to the disassociation of humans from their environment and led to selfish, desire-driven exploitation of others and the Earth's resources. Netzah struggles to liberate the prisoners of the senses through dreams while people sleep; to free them from their living nightmare.

It is important, then, that humans begin to open up more consciously to Venus and Netzah by stilling the mind and allowing the Earth and its web of life to again realign our consciousness with the greater plan of creation. As this process of dreaming potentizes the image base of humanity, the collective consciousness can begin to rise above Netzah on the waves of highest emotion, Love, to glimpse again the unifying Light of the Higher Mind, the soul, which is represented by the Sun and Tipareth.

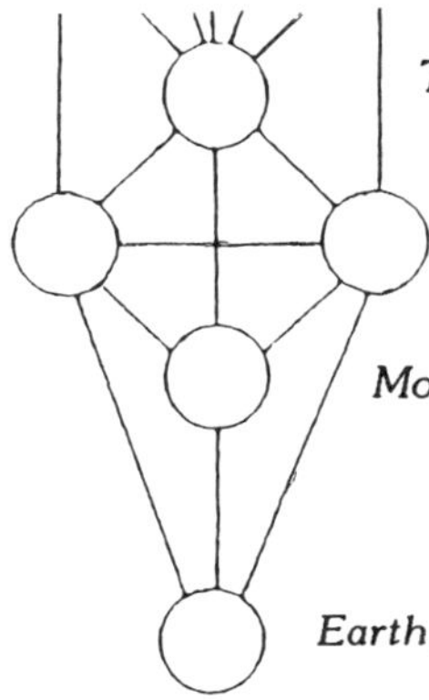

This rising of consciousness from Netzah to Tipareth liberates the being from operating solely in the world of substance and actions. Once Higher Manas, soul-mind, is active, consciousness then becomes free to perceive the world of formation and work with the laws that bring consciousness in and out of manifestation.

Tipareth is connected to every other sphere in the Tree of Life and is thus the mediating point of intelligence which relates the higher spheres to the lower. In humans this is the soul which relates spirit to body. Once the solar fire is alight, the unification process begins, matching body to spirit.

The natural pathway of ascent along the Lightning Flash leads from Netzah to Tipareth, from Venus to the Sun. By climbing this rung on the ladder of Lights humanity is introduced to the Higher Self. This pathway between Sun and Venus is symbolically represented by the Death Card in the Tarot.

The passions and egotistical desires of the limited self must die so that the true unity of the soul may illuminate the being. The individual needs to balance the conflicting influences of reason and passion and this can only be achieved by denouncing both and rising into a plane of pure, unattached perception. Self is no longer the servant of the body and the senses, but becomes aware of the higher purpose that serves the evolution of the One Light that is in all things. Hence the previous limited definition of Self must die as the radiant Self is realized.

It is interesting that this path is also ascribed to the Egyptian deity Kephra, the beauty and mysterious secret of the dawn. Venus, the morning star, rises and precedes the sun's glory. While the sky is still dark, Venus heralds the coming of morning just as Venus leads human consciousness from darkness to Light. Kephra is thus the initiator into the mysteries of Venus and passage into the higher solar realm. It is Kephra who has revealed the mystery that is written here.

Ritual of the Star

The symbolic connection between Sun and Venus occurs astrologically with the conjunction cycle. We saw previously that this cycle of ten conjunctions repeats itself each eight years and forms a pentagram within the heavens. The moments of conjunction are therefore very potent times for connecting the collective soul of humanity with the body, Venus to Earth, through ritual performance. These magical rites, dedicated to the Goddess as a mystical marriage of Heaven and Earth, are very powerful in uniting our global mind with its Higher Self. In chapter seven we will look at some possible ways in which the spiritual dimensions of this rite can be expressed in practical forms, either individually of in groups.

Venus in Astrology

We have already seen the significance of Venus as hierarchical ruler of the zodiac sign Capricorn. Venus is also the esoteric ruler of Gemini and orthodox ruler of both Libra and Taurus. Her influence in astrology is closely allied with the nature of the Goddess - magnetic, emotional and receptive. Traditionally, Venus has been connected with beauty, sensual pleasure, art, music, nature, fertility, love, marriage and the feminine. A well placed Venus in the horoscope would bring success and fulfilment in these areas while a badly placed or afflicted Venus position would bring demise. Her primary principle is magnetism; the ability to attract people, possessions or situations. The purpose of Venus is to bring about union, to merge self with non-self.

The symbol for Venus is the cross surmounted by the circle. The cross represents spirit's intersection with matter while the circle represents the unity which is above the sacrifice of light in form. This symbol is directly connected to the symbol for the Earth, again demonstrating the close affinity between these two twin soul planets.

Symbol for the Earth

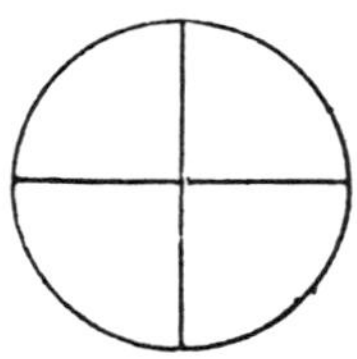

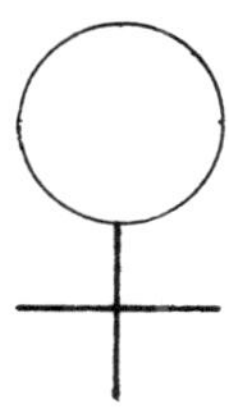

Symbol for Venus

Here the cross is enclosed in the circle which represents the containment of spirit in matter. Venus, on the other hand, offers an openness for upward evolution as opposed to the karmic bind of the 'word made flesh' represented in the symbol for the Earth.

Venus in Gemini

Venus is the esoteric ruler of Gemini. Gemini is the constellation through which love enters our solar system and Venus, planet of love, is the principle sphere to impart this attractive force. These combined influences cause opposites to be attracted to each other and particularly spirit to matter. Love is the gateway to higher knowledge and the soul. Venus and Gemini ensure a constant active interplay between the pairs of opposites, leading to eventual synthesis and union. It is this union of spirit and body which brings a new identity - at one with the higher mind and universal forces.

Venus in Libra

Venus is the source of intelligent mind and in Libra, the sign which Venus rules from the orthodox astrological view, this intelligence mediates between personal desire and spiritual love. Venus is the intermediary between higher mind, Buddhi Manas and lower mind, Kama Manas. The Libran influence is one of balance between these two aspects. In our mundane world we see the Venusian qualities of Libra as a feminine, soft grace and artistic beauty that balances the hard, masculine thrust of our aggressive world. She stands halfway between Heaven and Earth, balancing the two and ensuring that Light serves the dark below.

Venus in Taurus

Venus is also the orthodox ruler of Taurus, an Earth sign. Taurus has long been linked with desire and Venus in this place brings intelligence into desire. Thus the emphasis becomes one of searching for value and meaning in life that enriches and satisfies spiritual as well as physical desire. The intelligence engendered through Venus allows the presence of spirit to be identified within the simple forms, thus enriching the quality of Life. It is this innate appreciation of the art and beauty in the everyday forms of life and nature that characterizes the relationship of Venus and Taurus. Love has its dual aspects in desire for physical relationship and spiritual love. The lessons of Taurus allow desire to play out its drives while Venus continually refines the quality of experience sought.

Venus exalted in Pisces

Venus is exalted, meaning that its powers are enhanced, while in the sign of Pisces. This is because the Love principle reaches its highest evolutionary development in this sign. There is deep sympathy and compassion and an innate feeling for the essential unity of all things with Venus in this sign. Empathy combines with deep emotion allowing the attractive powers of Venus opportunity to merge with the spiritual evolution of humanity. Thus the ability to serve others is enhanced by Venus in Pisces.

Venus in detriment in Scorpio

Venus is in detriment, or has its qualities diminished, in the sign of Scorpio. This is because the intense desire-nature and sensual tendencies of Scorpio tend to bring the lower aspects of Venus - physical desire and preoccupation with sex - into dominance.

The soft, sensitive qualities of Venus are hardened by the male drive and aggressive intensity of this sign. Venus in Scorpio leads to a deep purging of the sexual and emotional nature, bringing purification through trial and adversity. Eventually the intelligence aspect will rise into dominance over desire and idealism and personal sacrifice will become more prevalent.

6

Planet Art

"As she stood without moving, it seemed to Kassandra that the air was filled with faint sparkles. as if the veil between the worlds grew thinner. Helen, touched with that golden shimmer, seemed to gain height and majesty and glow from within with beauty beyond anything human. It was no woman but the Goddess herself standing on the wall."

'Firebrand'
Marion Bradley
Page 370

The Earth is ready for a renaissance in Art, guided by the benevolent Goddess of beauty and creative inspiration, Venus. As Her influence waxes with the birth cycle of Aphrodite, symbolically relived by the motion of the planets, a new celebration of excellence and awe will emerge to guide the culture of the planet. What has now begun is the birth of a Planet Art, uniting humanity and the Earth in a spiritual bond, transcending the limitations of the previous age.

The relative health of any civilization can be monitored by the quality and abundance of its Art. When a society loses its creative inspiration it stagnates and degenerates. The human spirit must be kept alive with pageants and ceremony, music and theatre, to uplift the view of the common people; to demonstrate the invisible holy connection between all humans and between humanity and the Earth. Left to their own devices people are debased by the meanness of commerce and greed, lust and debauchery. Art and more specifically religious Art, has served to cast a net into the waters of the soul, bringing to the surface the higher qualities of humanity. Through theatre, music and the

Arts the great images of the culture can be brought to life, drawing on the unseen influences of the great souls who have preceded us, and for a moment, at least, bring Heaven to Earth.

Art is more than pretty pictures, music and dance. Art is magic. Art is the living expression of deity. Art is the soul in triumph. When the Gods dream, they dream Art. Art is innate within us, needing no other reason to be, other than the will for pure self-expression. Art is healing, enriching and stimulating. Art is beauty, harmony and grace that continues the creators work.

Our present world is crucifying Art in worshipping the false Gods of economics, science and technology. The intellect has become more than the logistical tool that it was intended to be and now suppresses the emotional, psychic, imaginative and spiritual roots of Art within humanity. Economics rules human creation so that all forms become utilitarian, mass-produced and usually square or rectangular units, to be used once and disposed of in ways that trample the unique, life giving Art forms of the greatest Artist. Human life has become devoid of inspiration and imagination. Artists survive only in context to the great beast, devoured in the soul enslaving pastimes of advertising agencies and bound to worship the twin gods of wealth and consumerism. Even the more independent artists, the rock musicians, are bound by contract to the cult of popularism, allowed only to record songs that conform to an aging formula, ensuring financial success triumphs over pure Art. Our artists have reflected the black pit we have fallen into, through such self-destructiive mediums as punk-rock. Anger has replaced beauty as the main inspiration of Art in our culture, with those artists who still possess a sense of harmony and balanced proportion relegated to rural retreat, without patron or church to aid their soul's triumph. Technology has enslaved us to duties that keep machines running. Mining machines devour the Earth, saw mills and woodchip mills devour forests and fishing machines destroy life in our oceans. Our machine servants have become our masters, proliferating across the globe, destroying the oldest and richest artistic cultures and turning simple crafts-people into consumers of technology's harvest.

As environmental destruction and the fragmentation of societies occurs throughout the world a compensation in the human spirit is inevitable. The balancing impulse is Planet Art. We have created the evolutionary pressure that is necessary for humanity to become truly one global people. The Earth, our common ancestor, mother and nurturer is approaching a crisis which we cannot ignore. Planet Art inspires humanity to begin the unifying movement, the Olin Yoliztli, movement of the sacred heart, which is an innate recognition of our common bond with the heart of the Earth.

Our technology has spawned a global electronic communications system which, when employed for social unity and education of the higher spirit of humanity, will be instrumental in uniting us in a global harmonic. Art, inspired by the Earth and her needs, will reach a new pinnacle as the healing renaissance finds strength and expression. Individual artists will resonate with an emerging planetary impulse, intuitively expressing a common theme which will connect artists throughout the world. We are now (1987-1990) seeing the preparation for Planet Art to rise into the consciousness of humanity as a real possibility for future direction. The vision necessary to change the current path of destruction of life-support systems has been seeded into the global mind. Planet Art, as it emerges from the decaying old paradigm of economic growth at the expense of nature, is set to mobilize the will of global humanity. A positive message of planetary repair and spiritual unification, not in a religious sense but in the uniting of the spirit in a higher purpose, will captivate the enthusiasm and empower a joyful relief from previous ages lived beneath the gloom of self destruction. The Earth will reveal her greater purpose through the human kingdom. By the end of 1993 the strategies for planetary purification will be well established and Art will flower, aided by global communications.

Nature and Magnetism

Planet Art begins as an intuitive assessment of changes in the subtle energy fields of the Earth and Nature. Art in this sense is more than an inspired feedback of sensory impressions from the environment, it is a total attunement and action which enhances the creative and magical nature of Life. The Earth communicates to humans through Her magnetic field. The aesthetic quality of Earth, which nurtures the psychic and etheric body of humans, is described by the ancient Earth-based cultures that developed geomancy. The Planet Art Network (see references to Jose Arguelles' work in the bibliography) is an alignment of human senses to the magnetic currents that describe the Earth's neurological membrane. As humans align to this organic network of sacred sites, inter-connected by lines of bio-magnetic force, a harmonic alignment of human mind and planetary mind is achieved. Thus the 'Sacred Sites Meditations' that have proliferated during the preparation cycle (1987-1990) such as Harmonic Convergence, Earth Link and others are instrumental in aligning human consciousness to the thinking membrane of the planet. The end effect of this alignment is an organized network of humans prepared to implement the planetary will.

Venus is directly related to nature and, through the attractive power of Love, to the phenomena of magnetism. This is why our present emphasis on the Morning Star as the mystery beacon for initiating us into the energies of this coming age is so important. We have forgotten how to love, connect with nature and feel the subtle magnetism of Earth-thought. If we are to heal the Life membrane of the Earth, and we must if we are to survive, then Venus is the primary energy and source of spiritual intelligence which can aid us now. She can help us re-attune to the spiritual body of our Mother and help us to enact the Earth's greatest needs.

Whenever we find reference to Venus in one of her guises as the Goddess, we will discover a culture which honoured the Earth and maintained a mystery school to teach about magnetism and nature. As the Goddess re-emerges as the balancing cosmic influence for humanity, Venus, in her many forms, will again become an overshadowing influence in spiritual ritual and devotion. Venus will restore the feminine virtue of receptivity so that we can respond to the evolutionary renewal of this coming solar age without resistance. Her attractive power fills the hearts of the world and bids them to welcome the bride groom, the new Jerusalem, for the next evolutionary cycle. Her heart beats within the pulsing web of Life and animates the spirit of Nature, asking us to breathe and accept our place as part of her greater Life and purpose.

Learning from Nature

All life is connected in a web of relationship, for all life is part of the ONE life. The face of the natural world is the face of the creator. Every rock, tree and stream bears the imprint of the perfect intelligence. The virgin landscapes of our world shine and glisten with the FORCE that issues from the stream of life itself.

As consciousness rises from the world of effects into the world of causes, the unity of all living things becomes an experience of unquestionable and absolute reality. Everything is conscious and active. Everything has purpose and intelligence, because everything shares in the Divine nature of the ONE. The consciousness of humanity has strayed far from the inner harmony of being. Like a loose string in a finely tuned lute, the song of human endeavour sounds forth as a discordant tone amid the chorus of the angels. Little does the isolated human personality register of the unified presence, which laps like waves upon the shores of the island of self.

Yet each wave that breaks upon the shores of the senses is one in a harmonic series. In the words of the popular artist Sting "Love is the seventh wave", the recurring harmonic pattern of the living universe. Rhythm is the key underlying factor within all manifested things. Molecules vibrate, the spheres sing, the heart beats and the breath flows in and out. Vibration is the law of the universe, yet the human animal does not know which of the many notes to sing.

The sixteenth century Alchemist, Paracelsus, whose knowledge of medicine enabled him to cure entire villages of the plague, declared that "nature was his greatest teacher". He wrote that "If we know nature, we know man and if we know man, we know nature". The laws of the Goddess are the laws of nature. The mysteries of existence are the mysteries of the winds, the seas, the forests and all the phenomena of the natural world.

The cities and factories of the human world operate in disharmony to the balance of nature. The mass organism of humanity is devouring the wild and free spirit of nature at an alarming rate. What will be our fate when the forests are all gone, the rivers all poisoned and soil stripped of its nutriment?

By attuning one's self to nature it is possible to discover one's own true nature. By retreating into the wild places and meditating with the trees, the birds and the wild spirit of nature the revelation of the wild, free spirit within becomes known. The continuous reflection of the peaceful, honest, pure being exhibited in nature brings forth the honest purity of one's inner being. A tree does not pretend to be a rock, nor does a bird pretend to be a spider. Everything in the wilderness is what it is without pretension. As one absorbs the impression of the total integrity of the elements of nature, the false assumptions and conditional views of reality fall away one by one. Eventually, after many days spent alone in the forest, the spirit of one's true being may be realized. One may then dance naked in the sunlight to the beat of one's heart. Then may the note of the inner life sound forth in harmony with the total OM of creation.

In contrast, human misconception views nature as being hostile. The wild enemy must be conquered and tamed, lest it overtake the cultivated land. Silence and stillness are uncomfortable for the human animal, for a true reflection of the wild, free spirit within threatens to overrun the cultivated fields of the false idea of Self. Material humanity fills life with constant, nervous activity. Desires become appetites and then addictions to unnatural ways and harmful activity. Yet the trees forever long to share a quiet moment and the winds whisper the secrets of life to those who have the patience to listen.

Amid the cooling green of the forests deep and restful, a soul can stretch and breathe. The song of the wild woods is a melody that enchants the senses with the Living Light of Life.

Beneath the veil of green the trees of Life serve the way of the Mother. In peace they grow and give fruit and sustenance to the creatures below. Love of Life have they who are the leafy ones of the silent glade. Each has soul alight with the shining silver of the Moon. Invisible light seen by the wise ones of every age.

Perfect light of love Divine
Curling creeper climbing vine
The hound of heaven's triple ray
Rustle leaves, branches sway
Seeds of sun in virgin Moon
Songs of woods, forest tune
Pulse of love unties the twins
Trees of life, Dolphin fins
By the old one's rainbow rings
By Morning Star the forest sings

Every tree is an intelligent being, focussing the pure cosmic prana of life into the world directly from the source on the Buddhic plane. Trees are the living bridge between the Buddhic and Kama Manasic planes corresponding to the soul and lower personality in humans. Most of the vegetable kingdom came into creation during the Venusian ruled Lumerian era. A bridge between spirit and matter, vegetation acts as a gateway for Love vibrations to nourish the rest of Life.

Attunement of the mind with the consciousness of the tree allows recognition of the inner pathway by which Love flows into ones being. Certain old mature trees in the forest embody a guardian spirit, which focuses the distribution of cosmic prana within a given region. Through their age and maturity these forest keepers have undergone a cycle of initiation which allows them to potently ensure that the forces of life energy flowing from cosmic source are correctly balanced within their sphere of influence. When the forests are stripped of the living spirit of this wise wood by human interference due to logging and clearing, the supply of cosmic prana is diminished and the equilibrium of life force in the region suffers greatly. Immature trees, separated from the living teachings of their immediate elders, do not develop onto higher initiation levels without the assistance of attuned forest keepers, such as the Druids, who are able to create consciousness pathways from other regions which permit "the whispering of the wise woods" to flow and enrich the lives of aspiring forest giants.

In many places the spirit of the woods has been broken and trees that remain are lost without direction and purpose. Diseases, die-back and loss of regenerative power are the physical symptoms of a broken spirit. The few remaining strongholds of wisdom within the forest kingdom are under assault from the destructive forces of human greed and ignorance. The guardian spirits of these remaining forest regions have little strength left for teaching outside their own regions as all their energy is focussed upon survival and defence. Without the protection and assistance of human lives the entire bio-system of the Earth, Gaia, will perish.

The Guardian-keepers, the old trees and the wise wood, seek the assistance of human souls at this time to aid in the reversal of the present decline in the life-energy of the planet. They are ready to teach and show those human lives who have awareness of the Tree of Life within them the secret ways of nature. By breathing and circulating light through the guardian trees the mind may be opened to perceive invisible currents flowing between oneself and the life of the tree.

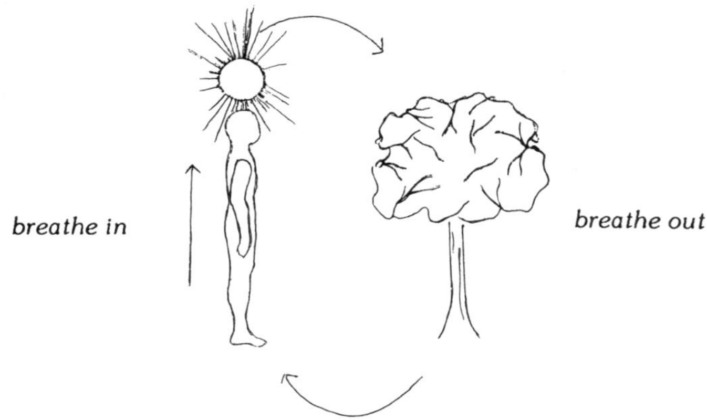

To the inner ear, the tree may speak its peace and be known as a living vegetable soul. By attunement to the tree one may learn Love, harmony and devotion.

A lesson may also be learned from the so called primitive cultures that preceded the present swarming of industrialism. The American Indians and the Australian Aboriginals both revered the land and held it to be sacred. Every leaf, tree, river and mountain was sacred to these people. They did not abuse the land but lived in harmony with all of the natural elements. The realization of the sacred nature within all things is the revelation of the Shekinah, the divine presence. If we can not see the presence within the natural forms of the wilds, how is it possible for it to be revealed in the concrete canyons of the city?

If your heart is pure and you maintain a quiet, peaceful determination to contact the invisible spirit of nature, your efforts will be rewarded.

The Face of the Goddess in Nature

The early Earth religions worshipped the Goddess and held reverence for her temple, the sacred groves and springs where her presence is strongest. Belili or Beltis, a Goddess of the Morning Star, was widely worshipped as a Goddess of wells, springs and trees, particularly the willow. She was originally a Sumerian Goddess and linked to Venus by the Sabians who used her name, Beltis, for the planet Venus. She is the 'Queen of Spring' and 'Mother of the Willow', in whose honour willow branches were waved at the spring equinox, particularly before sunrise when the Morning Star was rising in the East.

The connection between Nature, trees and the planet Venus was embodied in the sacred rites of the Goddess. Her rituals were always carried out in a natural setting rather than a church or other man-made structure. The purpose of these rites was the renewal of the energies of Life, within the community of humans and for nature as well. The connection made by trees to the cosmic flow of Prana was well recognized and incorporated into the rituals. Yet it is the human spirit which makes the bridge between the inflowing cosmic prana and the Earth's mind and being. The rites within the sacred grove were just as important for the renewal of the Earth as for the renewal of humanity.

Everywhere, in every culture, traces of a universal sacred Earth science exist, indicating that throughout the world people observed rituals of seasonal and astronomical determination, in which an unseen celestial influence was fused with the terrestrial currents to enrich and make fertile the soul of the planetary being. Stellar and planetary observances were of paramount importance to this universal science as is evidenced by the star shafts of the Great Pyramid, the alignment of standing stones to mark sunrise at the solstices and equinoxes and in the calibrations of celestial influence on the Chinese geomancer's compass.

Yet the knowledge of this sacred science began to decline as the Goddess was usurped by her sons and fathers as the dominant religious deity. The erosion of knowledge appears to have started about 4,000 years ago, as the Atlantean roots of wisdom became diluted following the destruction of Atlantis. Because the interaction between human consciousness, the celestial energy and the terrestrial currents occurs as a magnetic exchange, there has been a subsequent decline in the strength of the Earth's magnetic field as a consequence of the demise of geo-harmonic ritual. There is also evidence that much of the world that is now arid was once luxuriant forest, and that the diminishing strength of the total life force of the planet caused by deforestation is reflected in the waning magnetic field. According to Dr Hodder of the Institute of Geological Sciences at Edinburgh, the present decline in the magnetic field is in the order of three percent per century. After being stable for 700,000 years a decrease began 4,000 years ago and the total strength of the Earth's magnetism has decreased by half since then. If the trend continues the Earth's magnetic field will disappear completely by the year 3401!

Magnetism is essential to life and part of the mystic domain of the Goddess. She is the magnetism of the womb, the creative imagination that attracts spirit into life. Originally she was the possessor of all the mysteries, the Great Mother who brought all Life into existence. Gradually, however, the human spirit has become obsessed with the gateway into death, rather than entry into life. The religions of the Father God, which have gradually replaced devotion to the Mother, largely concentrate upon the afterlife, maintaining that Earthly existence is totally corrupted rather than sacred. As the ideologies of her consort and then usurper have come to dominate the spiritual attention of humanity, death has overshadowed Life as a gateway to the central mystery. It is not surprising then to see the Life force of the planet diminishing as the net power of human thought condemns life as inherently sinful. Thought is the main creative power on Earth and we will manifest the reality we strive for. In realizing this we have a tremendous opportunity to balance the opposing, yet complimentary forces.

The awakening of the Goddess within is the natural balancing influence necessary for humans to again come in harmony with the natural environment. That I have focussed primarily upon the Goddess in this work is not to put down the God, but rather restore him as a loving partner, rather than a cruel tyrant. If Love is to succeed, and this is the purpose of Venus, final Victory, than neither male nor female can dominate. There is room upon the Earth for all people if we establish a proper perspective relative to our environment. That perspective, however, can only be obtained by balancing the aggressive power of technology fuelled by science with feeling and respect for all Life, not just our own species or immediate family. Let the Pagans become a bit Christian and the Christians a bit Pagan. Let the feminists love their husbands and help them understand the intuitive and feeling aspects of their nature so that they in turn can love and respect woman. The initial process of shedding the yoke of oppression is in motion, particularly in Western societies, where women now command much of the same respect and rights in society as men. When the revolution has been successfully accomplished then it is time to reconcile divisions and strengthen the bonds of peaceful cooperation. The world is waking up to the waxing power of the Goddess renewed. We need Her powerful re-alignment of human experience to overcome our one-eyed path of selfish destruction. The woman knits the family together with a bond of Love. Humanity is one family and the Goddess is our Mother.

In the Mothers Womb

The Earth is our Mother, the womb of Life. This is not an intellectual assumption or a poetic gesture, it is an experience of Truth. Native Americans have a ritual where they spend the night alone buried in the Earth with a blanket over them, facing the fears of the naked night. In this ritual the separation of Self from the environment dies as one is reborn from the womb of the Mother at dawn. Similar practices existed throughout the world as the mysteries of the cave, the dark place within the Mother.

Much of the suppression of feminine nature arises out of a fear of darkness. The little self cannot bear the challenge of facing its hidden fears and so condemns those who bring attention to these fears. The night of fear ritual is essential in liberating the soul from the net of poison that keeps the personality bound in a ring of fear. Fear must be challenged and the strength to meet this challenge comes from the Earth herself. Each soul is part of Her thinking membrane, a latent seed of creativity and magic. When we accept that we are indeed one with the mother and offer ourselves up to be reborn in her womb then we become a child of the planet, free to be brave and do Her will. In Egypt, initiation took place in the crypt of the Great Pyramid, with the aspirant left over night in the stone coffin which was lidded and sealed. Fear becomes power once it is faced and overcome. This is what the Dark Goddess teaches us, to trade fear for power.

The Australian Aborigines also held the Mother in great reverence. The following passage from an unknown source demonstrates the roots of their culture in the Earth:

> "Birds, animals, reptiles, water, sky
> We belong to the ground
> It is our power and we must stay close to it
> or maybe we will get lost
> If you do not stay close to the ground
> you don't know what you are anymore".

As the feminine Ray advances upon the collective psyche of humankind, a more connective relationship with the Earth is emerging. We are beginning to let ourselves **be** and surrender to the Earth. When a child is distraught it turns to its mother and surrenders all its feelings to her nurturing power. The stress and tension of modern living does not allow us to show this 'weakness' beyond childhood. Our feelings well up inside ourselves unexpressed and unrelieved and block our sensitivity to others and our emotions. Families have become nuclear and unsupportive of our real emotional needs and there is nowhere to surrender our loads of anguish ... except to the Earth. The Earth can absorb all of our pain and tears and provide the love and acceptance that we need. This is what happens when we let go, surrendering to the Earth, Her Love and will. As more and more of us identify with the planetary Being, seeing our lives only in context to the larger picture of Life and Life creation, the transformation that will take place will be astonishing. War will cease between nations when we realize that the real war we are fighting is with the death-wish for self and the Earth. We cannot contain our pain and live a healthy, happy life. No matter how proud we are we need to show weakness and admit, at least to ourselves, that we are fallible and often imperfect creatures. The tears of such admissions are tears of joy for our Mother. Our strength is in our humility to surrender our egos and let ourselves **be**, without self expectation or preconceived airs. When the whole race admits its faults and cries, then will the Mother enfold Her children and the Love of the planet be shown to us. Individually the experience of surrender can occur at any moment. One practice for Earth surrender and attunement is to visualize yourself falling into the Earth; entering deep into Her core. As you descend deeper feel the layers of resistance strip off as you become more and more in tune with the planet rather than your separate identity. Feel yourself being absorbed into your Mother's heart, like a hurt child seeking refuge from a hard world. Feel the nurturing strength that the Mother gives you.

To truly surrender is a great mystery. When you enter into the heart of the Earth you will see Her mystery revealed. The Earth is pregnant. Venus has magnetised Her body with a sacred germ, an immaculate conception of cosmic proportions. The Earth is becoming sacred, filled with Light and giving birth to the star seed within, the living light of Her soul. Humanity is part of the transformation that leads to birth. Our crisis is the onset of labour. We are becoming part of a New Earth, a body of Light.

The Immaculate Conception

Venus empowers the creative imagination of humanity to aspire to perfection. Deep inside each person is a soul longing to express its ultimate desire - to bring spirit into manifestation. While the soul lies asleep and dormant, the individual feels a great emptiness and lack of purpose in life, which can gradually become a self-destructive frustration. Yet when the soul sparkles into life, a sense of beauty and aspiration arises from within. Each soul arrives with a perfect seed plan for life, a jewel of shining purpose which radiates an ideal of perfection. As Love bridges the gap between soul and personality, the forces of idealism grow in the individual, empowering a creative search for expression, so that the inner perfection finds root in the outer world. At the pinnacle of this search for meaning is the immaculate concept, the shining soul-star in its complete unfoldment.

The vehicle for bringing the soul-star into radiant activity is the Dove of the Holy Spirit. In a previous chapter we traced the connection of the Dove and Holy Spirit to Venus. When Jesus was baptized by John, the Holy Spirit descended upon him as the Dove. When Mary was impregnated with the seed-life of the Christ, it was the Holy Spirit which again provided the vehicle:

> "fear not to take unto thee Mary thy wife: for that which is conceived in her is of the Holy Ghost."
>
> *'Matthew'* 1:20

The Holy Spirit is the feminine, magnetic quality of Godhead, the third aspect of the trinity or Mother aspect. Father and Son are sterile as it is only the Mother who can draw life out of herself. The Holy Spirit is the energy which quickens the evolution of the soul. It is the womb from which the perfected self is born. Spirit is within all things, all the time. Yet it is hidden. The Mother or Goddess is empowered to set the spirit free for She is the same spirit. The Father is Light, the Son is the Word and the Mother is Spirit which is the source of all substance. The immaculate concept is the experience of self as that very same maternal spirit.

The Goddess Returns

The weight if evidence is before you - the Goddess is returning. Though she has been ever-present, providing the very substance of all our lives, She is returning into our full consciousness so we can again perceive Her glory. Her mission is that of redemption, to free us from the false emanations. Her influence is the great resource which can realign our lives to harmonize with the Earth, nature and each other. Her many forms are one form, one Spirit of everlasting Life. She is our Mother, our sister, our lover. Her perfect being is revealed to us in the Rising of the Morning Star. She emanates our Victory, our triumph over matter. She is the fire that lights the soul. Blessed are we that She has returned.

7

Ritual of Love

There is a world of Light and Dark. Rest I, in the shadow ... of His wings ... 'Morning Star', last Light before the Dawn. It is my place of purpose ... the tender shadow ... the place of balance. Enter I, into the dark to bring light; bold and full of courage to face my fears; to heal my spirit ... to find my path behold ... VISION

Morning Star'

As the Goddess comes into full manifestation, there are certain times when Her rites need to be performed. Venus and Her orbital cycles are of paramount importance to the transfer of feminine power from sphere to sphere. At the times of Her conjunction with the Sun, Her seed of Light is volatile and present with us. When She rises before the Sun, Her illuminating presence heralds the dawn of our souls which is soon to come. When the Morning Star rises with the Dove (Pleiades), the Holy Spirit is revealed in the priestess and Her blessing is upon all within Her circle. Most of all when Venus rises when Neptune and Uranus are conjunct, Aphrodite will return to Earth and guide us in Love, Her art.

The purpose of Ritual

Ritual allows us to live our deepest meaning, to draw archetypes into consciousness and experience the presence and power of the Gods and Goddesses. Humans are an essential medium for transferring cosmic essence into the Earth. We are the thinking creative membrane of our Planet and responsible for the healthful influx of celestial currents into Her bio-magnetic sheath, thus increasing the vital energy available for all Life. For the individual, ritual ensures that spiritual understanding is derived from **experience** of energies and presence directly, rather than from just intellectual grasp. In the sacred circle the deities come alive and their power is channelled by their Earthly representatives to enlighten and heal those present. Ritual programmes not just the conscious but the unconscious as well; recrystalizing even the DNA of every cell with the energy of divine experience.

The effects of ritual are, of course, dependent upon the quality and degree of attainment of the participants and their principles. The greater the sanctity in preparations and the deeper the commitment of **all** the people participating, the greater the wealth of transcendent experience which may be gleaned from the sacred play. Ritual is divine theatre, at times spontaneous and playful, at others disciplined and ceremonious. There are no spectators, only participants, and even the sky, trees and the Earth are part of the play.

Life abounds with ritual, not always sacred but always rhythmic. This is the reason why rituals are performed at prescribed times to coincide with the rhythmic motion of celestial bodies. Music is thus an essential element in ritual. Rhythm and dance are integral to all cultures in their most sacred rituals. From the hymns of the Catholic to the drums of Zulus in Africa, music and rhythm are used to cultivate the inner sanctity or trance that precedes divine order and experience.

Age of the Seventh Ray

The Aquarian Age is governed by the violet or Seventh Ray of ceremonial order and magic. We are now in the transition stage when magic and ritual are being re-established as the dominant religious forms of the day. Along with the Seventh Ray, the Fifth Ray of 'Concrete Science' is also waxing into power. The Fifth Ray, as we saw in Chapter Five, is also the ray of Venus. The combined influence of these two rays will bring about an awakening of esoteric knowledge and its use in geo-harmonic ritual. Eventually we will see the emergence of a Planetary Art Network, which engages the support of millions of humans in creative ritual, combining esoteric craft with observance of planetary cycles to re-energize the Life membranes of the planet. All the gifts of the Goddess will be revered; music, art, nature, Love and Life-creation. This New Age will balance male and female as never before, allowing Truth to overshadow dogmatism and religious squabbles. Science, not in its separating analytical form, but as synthesis, the feminine, magnetic merging of truth into wholistic understanding, will vindicate spiritual truth and merge intellect with intuition, bringing greater tolerance and universal understanding to a uniting world people. All Gods and Goddesses will be seen to have a place in the play of divine energy. Even opposing religious views will be seen as two aspects of an inseparable whole. Information and celebration will become the dominant human pastimes as a wholistic world order dismantles unsustainable industrialism in favour of an Earth-friendly sustainable agriculture and communication based society.

Sacred Sites

The most effective Earth-healing rituals are performed at Sacred Sites. Sacred Sites are essentially places where there is a convergence of physical, magnetic and spiritual forces. This allows easy access and transfer of energy from one level to the next. The rituals outlined in this chapter are designed to instil the qualities and energies of the celestial Goddess into the Earth's life membrane. Sacred sites, particularly those which are more feminine in their nature like caves, springs and wells, are ideal places to conduct these rituals. Many sacred places are already widely known and marked, like the Great Pyramid or the Chalice Well in Glastonbury. Others, however, are waiting to be discovered. These can be found intuitively, or by dowsing, if sufficient skill exists. In my next book, *'Earth Service'*, I will explore in great detail the methods and techniques whereby you can learn to dowse and work with sacred sites. Until its release you will have to trust your existing skills in this regard. Some techniques were also given in my books, *'Earth Link 88'* and *'Winged Serpents Dancing'* which, unfortunately, have become collectors items. Your intuition is always your best guide.

When beginning to search for a suitable site ask the Goddess to help you. A suitable prayer may go like this.

Daughter of Light before the dawn
Sister of night whose Love is strong
I offer myself to your verdant throne
To serve your will on Earth and stone
Guide my way to your sacred vale
Where your presence is strong, vital and hale
May each step that I take embrace your way
Guide my soul through night and day
I ask you now by your holy name
Aphrodite, Venus who are the same

The feminine virtues of the Earth's landscape are soft, rounded and peaceful. There is an abundance of water and trees and forest. The Yin elements of water and Earth are strongly favoured. By rivers near soft rounded hills. By springs and in valleys. In quiet forest glades. Near waterfalls and in caves. By lakes and places where mists gather. Where landforms and trees reflect feminine anatomy. All these places are appropriate provided you feel the soft, nurturing presence of the Goddess with you.

Preparation for Ritual

The preparation and ritual described here is based on the Wiccan tradition of Witches. They may serve as a guideline for you in your ritual explorations, but do not feel bound to this form. Many other rites may be performed and I encourage you to look at alternatives and tailor a rite that best suits your needs.

The key aspect in preparation for rituals involving Venus and Aphrodite is the casting of a circle. Traditionally, according to Wiccan codes, the casting of a circle takes place immediately prior to the ritual. As the participants for the ritual gather, an informal sharing takes place, as each greets the other and catches up with their lives. A group purification should follow. A bowl of salt water is passed around the circle, generally in a clockwise direction, and each person in turn sprinkles themselves and the person to the left with a few drops from the bowl. Salt is a purifying substance and water is the element at the source of Life. This silent sprinkling is a blessing of oneself and the circle.

Next, the principle of the ritual, generally the priest or priestess, prepares the group consciousness with a guided meditation, linking hearts and minds together on a higher level of consciousness. Symbols appropriate to the purpose of the ritual work are recommended, with the opening of the heart a key to entering into the confidence of Venus. This preparation takes place outside the chosen site for the ritual. When everyone is psychically open after the meditation, the priestess, having chosen and consecrated a suitable magical weapon (for Aphrodite - a lamp or girdle) dips the weapon in the salt water and sweeps the weapon around in a clockwise circle, starting and finishing in the East. While casting the circle a prayer of protection is spoken which may go something like this:

I cast this circle with the Holy Names
Of Aphrodite and Venus who are the same
May Her power protect us, one and all
From evil and wickedness, hear my call
May we meet within Love, Truth and Joy
May the Goddess empower us as we employ
Our Spirits, Souls and Bodies to do her will
In this circle contain our power and skill
I consecrate thee in the Names
Of Ushanas, Muth, Aphrodite and Venus.

When the circle is cast an opening is made by lifting the weapon on one side, generally the North-East and the rest of the party files in and circles around. The entry way is then closed with a clockwise stroke of the weapon or 'athame'. Clockwise is a direction which binds energy together and is therefore appropriate for casting and entering the circle. Counter-clockwise or 'widdershins' is used in closing the circle and unbinding.

Invoking the Four Directions

Traditionally an invokation, or drawing down of the elements of the four cardinal points occurs once the circle is cast. A person is selected prior to the ritual to perform the invokation of each direction. Each bears a symbol representing the appropriate element. For Air, this may be a sword or feather; Fire, a candle, wand or incense; Water, a cup, chalice or thing of the sea such as a shell; and Earth, a flower, sheaths of grain, pentacle or fruit. These items are dipped in salt water and become athames or magical weapons for tracing the invoking pentagram. The standard invoking pentagram is the one which represents the Earth element. Starting in the East, in honour of the Morning Star, the Air priest or priestess faces this direction and traces a pentagram in front of them with their athame, beginning at the apex, bottom left, far right, far left, bottom right and returning to the apex again. While tracing this pentagram they might say:

I call upon the watchers from the East
To guard our rites and join our feast
All spirits and creatures of the air
Raphael, Sylphs and Eurus fair
Guard the Light and keep it strong
Share in our mirth, our joy and song

Next, the principle from the South repeats the tracing of the invoking pentagram using the fire athame saying:

From the South all watchers gather near
To cast out evil, hate and fear
With Holy Fire purify us all
Michael, Salamanders, hear our call
Notus guard us with your fire
Upon your flames take our Spirits higher

The water representative from the West then traces the invoking pentagram saying:

From the West come watchers now
To guard the gates, to keep a vow
Protect us now, hear our call
Gabriel, Zephyrus and Undines all
Purify with waters clear
Come gather round from far and near

Finally, facing North, the Earth principle traces an invoking pentagram with the Earth athame calling out to invisible spirits thus:

Come from Earth all watchers here
To guard this gate, to keep it clear
Uriel, Borus and Gnomes of the North
Protect this circle and all henceforth
From evil, wickedness and mal-intent
Maintain your guard 'til away your sent

Once these initial ceremonies are completed the circle is prepared for the ritual to begin. The above spoken words are my own, but other options are available in books such as *'The Witches' Goddess'* and *'The Womens Spirituality Book'* detailed in the bibliography. I encourage people to make their rituals their own by constructing their own prayers and invokations according to a general plan. There are, of course, many different ways of achieving the same result and it is up to you to choose a form that suits your needs. Truth is never yours until you make it your own by experiencing it.

Rituals, I believe, should not be performed by rote, utilizing the same words and actions each time according to ancient formulas.

Each moment is new, having unique potential for bringing through a new type of energy. Astrology is essential therefore in the construction of rituals appropriate for that particular moment. The positions of the planets in the signs and in relation to each other provides a potent analysis of consciousness potential. Certain symbols and forms of experience will be highlighted by the current potential described in the stars. As above, so below. Attention to this prime Hermetic axiom will ensure that your ritual is in tune with the potential of the moment.

The following rituals are offered as guidelines for events associated with the cycle of Venus. How you utilize this information is up to you and, by all means, elaborate and add to these suggestions with your own deepest inspiration. Truth is relative. Be responsible for your own destiny.

Ritual of Aphrodite

This ritual is based upon the Great Marriage symbolized by the Love and union of male and female. The Sun is the golden unifying force, traditionally masculine while Venus is the Love-Goddess. When they meet, at a star-point of the orbital pentagram, it is an ideal time to invoke the presence of Aphrodite in a celebration of Love. This ritual may also be used for other events in the cycle of the Morning Star, such as conjunctions with other major planets.

The ritual is principally concerned with the invokation of Aphrodite, whose presence overshadows the priestess. Though the full ceremony is described here with script, the priestess may take the course of the ritual along other lines if the Goddess is inspired to take charge. The atmosphere should be light and celebratory. Aphrodite **is** the Goddess of Love and pleasure, so ensure that there is plenty of room for mirth and frivolity in your proceedings.

After casting the circle and invoking the guardians from the four directions, the priestess is blessed by the priest. They stand facing each other in the centre of the circle by the altar, which is comprised of the athames and personal charged objects, crystals, etc. of the participants. The blessing can take a variety of forms, from the ritual kiss and blessing to the feet, knees, womb, breasts and lips with the appropriate words, or, perhaps, the priest may sprinkle the priestess with salt water, burn incense or draw the

Light down upon her with the raising and lowering of the hands. The form chosen will depend upon whether this is to be a hands-on ritual between intimate companions or a symbolic rite. I am sure the Goddess would approve if the man and woman were lovers and their intimacy was part of the celebration.

Once the blessing is completed, the priest invokes the presence of the Goddess:

I invoke thee Aphrodite
Queen of Love, golden & fair
Upon thy servant cast your presence
Bless her body, heart and hair
For your Love she is the altar
Centre of the sacred Life
Draw your spirit now upon her
Aphrodite come to life!

The high priest also requires blessing and the invoking of a suitable God-form. The priestess now blesses him in the name of Aphrodite and then invokes the presence of Eros thus:

I call upon thee Eros
Come forth with passion and flair
Your servant awaits your presence
I invoke thee now be there
Upon his blessed body
Cast your spirit, young and strong
For love will not be silent
Do not keep me waiting long

The two now rejoin the circle and lead a circle dance - preferably one in which men and women circle in opposite directions, exchanging partners after each round. In this way the direct energies of the God and Goddess are experienced by each of the participants. Once the circle has been completed, the priest calls upon the Goddess:

Hail Sweet Aphrodite!
Take my heart, my soul, my mind
I am yours for Love's great purpose
May our hearts and souls entwine
For Love you are the altar

Golden Light so fine
Take this servant for your purpose
Like the Morning Star you shine
Your rites are made for pleasure
Your Love like golden wine
I serve your beauty and your fragrance
Aphrodite I am thine

The priestess then stands forward and lets fall a veil or girdle as a sign of invitation and calls upon her lover:

I am Aphrodite
Golden and divine
By five pointed star I greet you
From within my heart it shines

The priest then steps forward and they embrace in the centre of the circle. The rest of the participants begin another circle dance and gradually, two by two, pair off and retire to complete the ritual in private. The priest and priestess remain until all have left and then retire to do the same.

While this ritual encourages the passionate aspect of Aphrodite, do not forget that this is only a veil before Her true Being. The Goddess, though She works through Love and passion, inspires the presence of the higher mind, imparting knowledge and wisdom. The sexual current must be focussed upward, like nectar into consciousness, rather than downward into generation.

Native American Prayer Ceremony

For those readers who require a more personal tribute to the Morning Star the following prayer ceremony may be appropriate. This rite is traditionally preceded by purification, fasting and sweat lodge.

> "I, identify myself to the Spirit World as Morning Star. It is the name for which I wrestled my fears and my past that I would be free from the arrows which pierced my heart. I will not go back to face my fears and past - they are dead as that old self is dead and I have risen as the Phoenix from the sand to new life. My path is now the flight of the Eagle to my home and his - the Morning Star. It is my choice, my place and my purpose.

I give now my ancient tribal identification so that all may know, above and below, I make these seven prayer ties by the Light of the Morning Star. I come with my requests and before the setting of the sun these prayers will come into being. So be it! It is done now in the physical and it is done in the spiritual.

1. I call forth the Spirit of the Wind. Come and bring Divine Love from the spirit world of the stars to the place wherein I stand - to this Sacred Wheel, to my heart and spirit, to my Mother, the Earth. Blow forth, O Mighty Wind of Spirit CHANGES. I feel your breath and accept your purpose. I make this prayer tie unto you.

2. I call forth now the Morning Star. Behold, it's Life, Strength, and Fruitfulness. The Morning Star ... clothed in red, the colour of Life and with the touch of the coming sun - you are shining in the heavens brighter and brighter. You are the breath of the new day. You never move and all the other stars walk around you. I call forth the Light and energy of this ancient star being. Talawasohu. Umpo Wacapke. The root and offspring of David - the bright and morning star - Eternal One. Come forth Aphrodite, the Goddess of Love and Venus. I make this prayer tie unto you.

3. I call forth now the door of the future and of destiny. Swing open, O sacred ancient door that we might pass through - Grandmothers and Grandfathers, Sisters and Brothers, Mothers and Fathers. I make this prayer tie unto you.

4. I call forth now spirit guides, ancestors, allies, totem spirits, Nunahe (the spirit of this place) ... those seen and unseen. Go forth now as agents of Light and Healing with my brother, the Wind. I make this prayer tie unto you.

5. I call forth the Keepers of the 4 directions and as I smoke the Sacred Pipe, I make this prayer tie unto you, O ancient ones.

6. I call forth those now that make their requests before the Light of the Morning Star. I make this prayer tie unto you and before the setting of this sun I will see them done in the physical and in the spiritual. So be it!

7. I call forth now Blessings and Healing. I call forth Vision, new reality, and growth. I call forth by the Light of the Morning Star spiritual awakening and courage. Go forth now on the spirit of the wind, my brother, and on the wings of the Morning Star. I make this sacred prayer tie unto you. So be it.

Go forth now Spirit of the Wind. Go forth Morning Star. Go forth Door of the Future and Destiny. Go forth those seen and unseen. Go forth the Keepers of the Four Directions. Go forth all you who make your requests by the Light of the Morning Star ... it is done this day before the setting of the sun. Go forth blessings, healing, vision, new reality, growth, spiritual awakening and courage. Morning Star sends you Forth! So be it!"

Ritual Written by:
'Morning Star'.

Meditation of Alchemical Marriage

Regardless of the type of ritual or observations that you make during the Venusian cycle, a technique of achieving internal marriage of your male and female aspects is an essential preparation. This meditation uses some basic principles of Craft to consciously generate alchemical fusion of Self with the anima or animus quickly and effectively. The result of conscientious practice of this technique is the blissful raising of the serpent energies and the opening of the spiritual heart. The blind and ignorant will, of course, read this and never actually complete the technique no matter how much I stress the importance of it or advise of the results. So be it! I can only encourage you to follow these instructions.

As in all meditation techniques, breath is the foundation. Begin by breathing rhythmically, measuring your natural, relaxed breath to ensure that the in-breath and the out-breath are kept constantly of the same duration. After five or ten minutes you will feel the energizing effect of gathering 'pranic' energy or Life force as a slight light-headed intoxication.

Using the full concentration of your mind and the power of each breath visualize an idealized figure of the opposite sex above your head, upright with their feet just above your crown. Take careful note of every detail of their being, from head to foot. Visualize them naked if you wish, for the raising of the serpent is both a sexual and spiritual rite. The men may invoke this Spirit of the Anima by vibrating the name of Aphrodite with each breath. The women may use the name Eros. When, and only when, the vision is complete and strong within you, proceed.

Now with each breath, at the crest of the in-breath, visualize a bright flame burning within the heart of the spirit above you. See and feel the flame intensify with each breath.

Next visualize a similar flame immediately above your head; between the feet of the being above. With each breath empower the flame in its rising intensity.

Now feel and see the flame begin to grow in your heart, again empowering it with the breath. Maintain awareness of the other flames as you do this, feeling their presence as well. When you can feel each flame with deep intensity and see them clearly within you and your higher form, proceed.

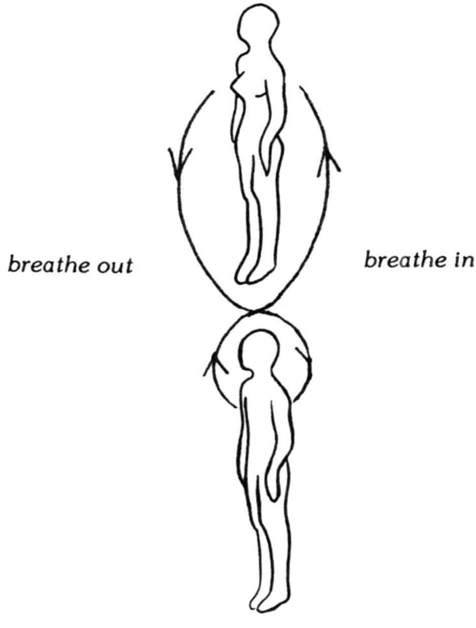

Meditation of the Alchemical Marriage

With the full strength of your concentration and awareness return your attention to the top flame - in the heart of your upper opposite form. As you breathe out, trace a pathway of flame from the heart above, down through your crown to your heart below using a figure '8' pathway. As you breathe in, trace a pathway of flame upward through each flame completing the figure '8' cycle. Be aware of the sensation of each flame as you pass your focussed consciousness through it. Keep the visualization of the spirit-double clear in your mind.

Continue this breathing pattern for as long as is necessary. I always find it entirely blissful and recommend it for any occasion particularly when the emotions are disturbed or you are feeling out of sorts.

You may also like to practice this meditation with a friend, preferably of the opposite sex. I this case you will need to visualize their idealized double above them as well. Back to back is recommended for platonic friendships while facing each other is much more intimate. In this case the breaths are synchronized so that you breathe in and out together. The figure '8' passes through your combined centres as illustrated below.

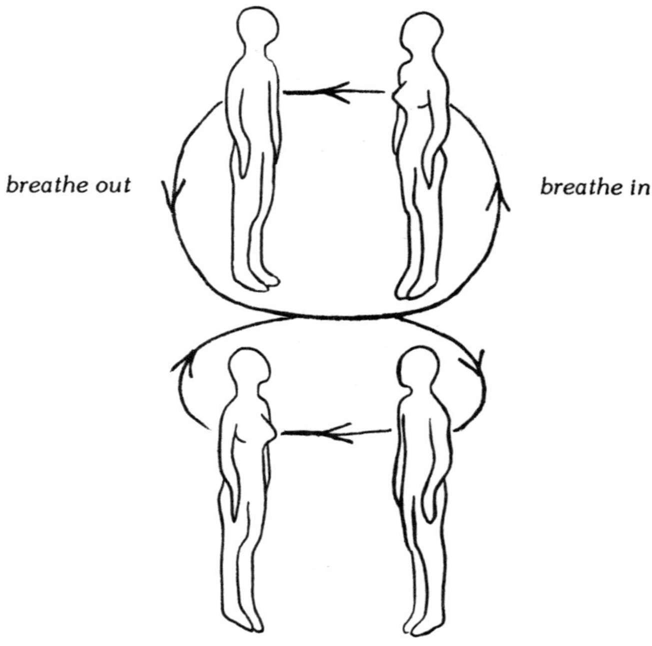

working with a partner

Invoking the Cosmic Pentagram

Venus Herself is the great magician and Her dance of the five points reveals magic that pervades the entire solar system. We honour her great rite by understanding the purpose of Her great cycle. I refer again to the pentagram described by the conjunction points of Venus and the Sun over an eight year cycle.

The pentagram we know as the most potent, holy and protective of symbols. It represents the higher mind ruling over the lower elements of being and the beginning of cosmic consciousness. The five pointed star inscribed within a circle is one of the oldest of symbols, having been traced back to pre-Babylonian Sumer. Inverted it represents spirit submerged and bonded with material form. Upright it represents the redemption of matter by Spirit which rules over it.

In magical ritual much use is made of the pentagram in the invoking and banishing of the elements. In its most common and widely used correspondence the pentagram has the elements attributed in the following order.

The invoking of various elements is very useful in drawing forth the qualities represented by the elements nature.

Air - stimulates the creative mind and intuitions
Fire - stimulates vitality and energetic expression
Water - stimulates the free flow of emotions and healing energy
Earth - stimulates practicality and responsibility
Spirit - stimulates the flow of cosmic energy and higher consciousness

The various elements are invoked by tracing the pentagrams according to the following directions.

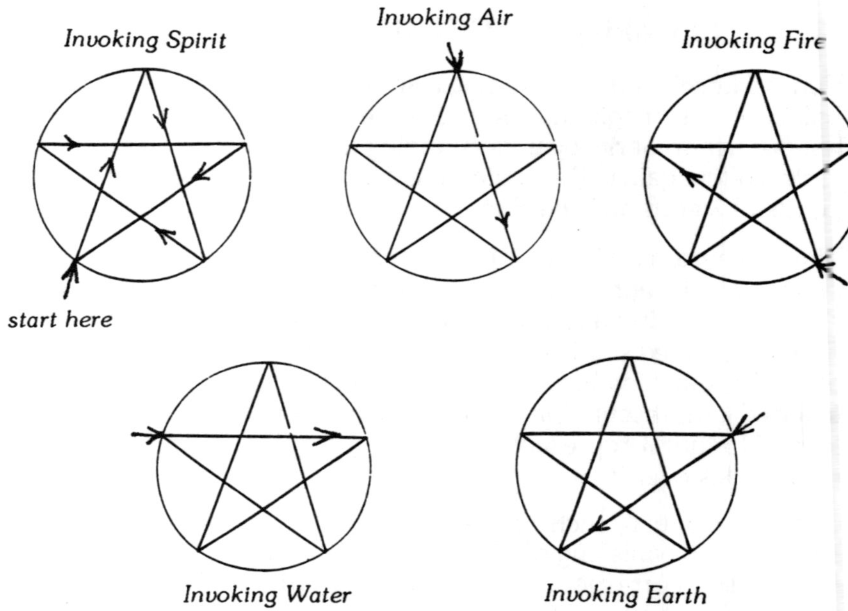

Banishing is useful for clearing excesses or unbalanced expressions of elemental energies. Banishing of the various elements takes place by tracing the pentagrams in the following diagram.

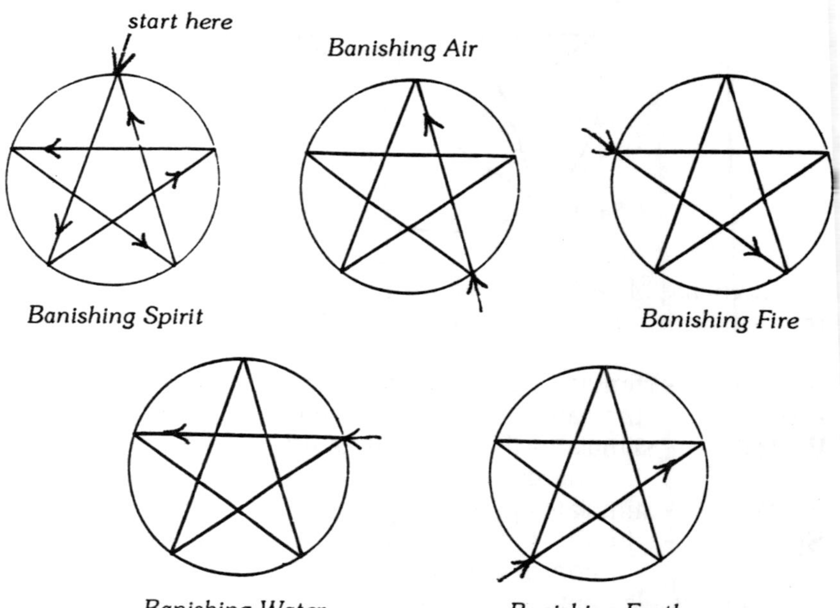

The directions given here for invoking and banishing may differ somewhat from other systems, including the Wiccan system used in the Aphrodite ritual earlier in this chapter. Many of the older systems utilize different elemental placements and did not account for the fifth element of spirit. This present system takes into account our progression into the fifth solar age and recognizes the spiritual activity of the fifth hierarchy. The system is based upon a natural harmony with invoking actions all flowing in the same direction and banishing in the opposite direction. This would seem to accord much better with the Venusian way of harmony within nature.

Various elements are represented at each of the star points during each of the ten conjunctions that complete the eight year cycle. These are determined by the astrological sign in which the conjunctions occur.

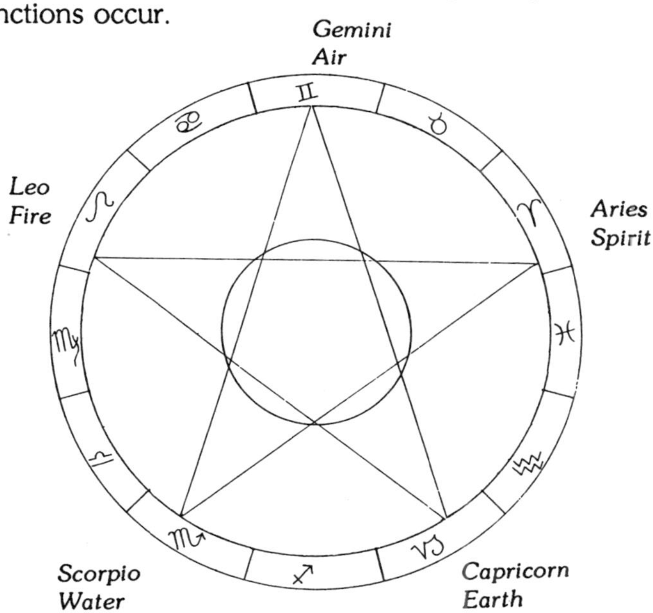

Ten sun/Venus conjunctions

In the present cycle which began on the 5th April, 1989 in the sign of Aries, which most represents spirit as electrical fire, the general direction of the elemental flow is in a counter-clockwise or widdershins direction. This corresponds to the banishing cycle of the elements and the releasing of energy from fixed structure. The Venusian cycle is a continual purification of the elemental energies.

As there are two conjunctions at each point of the pentagram, one superior and one inferior, a secondary effect ensues by observing the direction of transiting Venus at the time of each. When Venus is between the Earth and the Sun, She appears to be travelling backward through the zodiac. This retrograde motion means that Venus will actually be travelling clockwise through the pentagram at these conjunction points, thus inferring an invoking action rather than a banishing one. Therefore, there is a banishing and invoking conjunction at each elemental point during the eight year cycle. We can use this knowledge to create the appropriate pentagram ritual for each occasion in harmony with the great cycle. The following elemental forms are indicated on the following dates:

5 April, 1989 - banishing Spirit
19 Jan, 1990 - invoking Earth
2 Nov, 1990 - banishing Water
23 Aug, 1991 - invoking Fire
14 June, 1992 - banishing Air
1 April, 1993 - invoking Spirit
17 Jan, 1994 - banishing Earth
3 Nov, 1994 - invoking Water
21 Aug, 1995 - banishing Fire
11 June, 1996 - invoking Air

The cycle will continue in much the same pattern and can be determined at any time by noting the element of the sign in which the conjunction occurs. If Venus is retrograde the resulting ritual will be invoking; if it is direct then a banishing ritual is required. If you do not have access to a planetary ephemeris then try consulting your local astrologer.

Each element will infer a different emphasis upon the ritual. It is not within the scope of this work to provide the form for each of these rites. If the principles outlined here are followed through to generate your own forms then that is sufficient to ensure a basic harmony with Venus and Her cycle. The rest is up to you. Detailed astrological charts for these conjunction points are given in the appendix.

Birth of Aphrodite

In chapter five the parallels between the Greek myth concerning the birth of Aphrodite and our current astrological cycle were examined. The reasoned culmination of this mythological birth cycle will occur when Venus transits the Uranus/Neptune conjunction on January 13th, 1994. This date is an important emergence point for the rebalancing of human consciousness through the resurgence of the feminine spirit. Art will have reached a planetary climax as a new integrated spirit of unity and co-operation pervades planetary affairs. The all-important shift from industrialism to ecological communion will have reached the turning point in preparation for the Golden Age of Aphrodite.

The moment of Aphrodite's rebirth is a paramount moment for ritual observation. Please ensure that you honour this celebration of Love at a site sacred to you. The entire period from the 13th January to the Sun/Venus conjunction on the 17th January, 1994 should be a time of concerted efforts to renew the Aphrodite seed - the Love principle of the Holy Spirit - and ensure the complete return of wisdom - Sophia - for which the Goddess is known. The ritual principles given previously in this chapter may serve as a guide. Subscription to *'Earth Link Journal'* (see details at the end of this book) will ensure that you are aware of planetary preparations and up-to-date information as this magic moment approaches. Full astrological information concerning these dates appears in the appendix.

Venus conjunct the Pleiades

The moments when Venus is conjunct the Pleiades, constellation of the Dove and galactic source of the Holy Spirit, are times of the deepest reverence of the Goddess and her redeeming gift. The opportunity for the release of the in-dwelling spirit culminates during these times. As the Spirit is released, Her energies enlighten consciousness, heal and open the heart to Her cosmic Love.

The dates for these vital conjunctions during the next ten years are as follows:

24	June, 1990
12	April, 1991
25	May, 1992
5	July, 1993
25	April, 1994
10	June, 1995
3	April, 1996
10	May, 1997
24	June, 1998
12	April, 1999

Please note that the eight year orbital cycle of Venus repeats through these dates as well as for the Sun/Venus conjunction dates.

May the Love of the Goddess descend as a Dove upon you,
filling your Life with Her Holy Spirit!

Appendix A - Astrological Diagrams

Diagram One
Sun / Venus conjunction
5th April, 1989

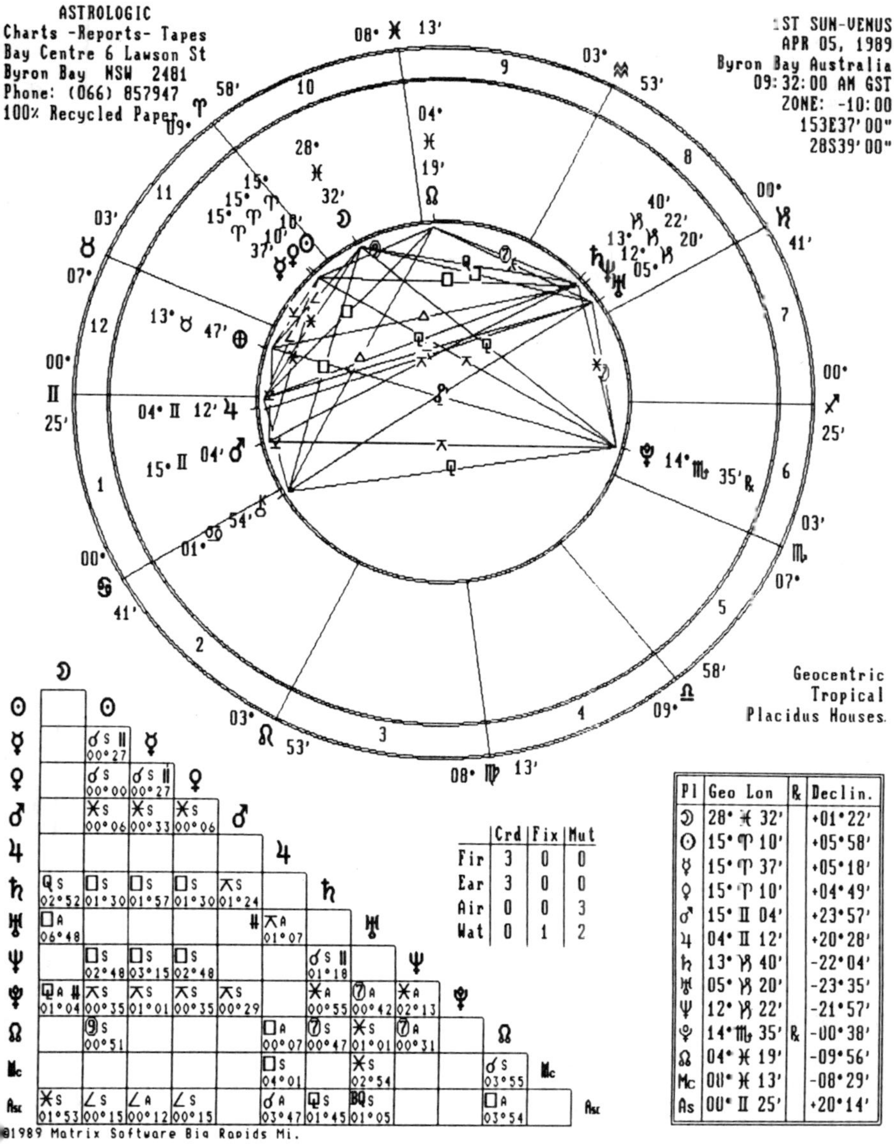

	☽	☉	☿	♀	♂	♃	♄	♅	♆	♇	☊	Mc
☉												
☿		☌ S ‖ 00°27										
♀		☌ S 00°00	☌ S ‖ 00°27									
♂		⚹ S 00°06	⚹ S 00°33	⚹ S 00°06								
♃												
♄	Q S 02°52	□ S 01°30	□ S 01°57	□ S 01°30	⚻ S 01°24							
♅	□ A 06°48				‖	⚻ A 01°07						
♆		□ S 02°48	□ S 03°15	□ S 02°48			☌ S ‖ 01°18					
♇	Q A ‖ 01°04	⚻ S 00°35	⚻ S 01°01	⚻ S 00°35	⚻ S 00°29		⚹ A 00°55	⑦ A 00°42	⚹ A 02°13			
☊		⑨ S 00°51				□ A 00°07	⑦ S 00°47	⚹ S 01°01	⑦ A 00°31			
Mc						□ S 04°01		⚹ S 02°54			☌ S 03°55	
Asc	⚹ S 01°53	∠ S 00°15	∠ A 00°12	∠ S 00°15		☌ A 03°47	Q S 01°45	BQ S 01°05			□ A 03°54	

	Crd	Fix	Mut
Fir	3	0	0
Ear	3	0	0
Air	0	0	3
Wat	0	1	2

Pl	Geo Lon	℞	Declin.
☽	28° ♓ 32'		+01°22'
☉	15° ♈ 10'		+05°58'
☿	15° ♈ 37'		+05°18'
♀	15° ♈ 10'		+04°49'
♂	15° ♊ 04'		+23°57'
♃	04° ♊ 12'		+20°28'
♄	13° ♑ 40'		-22°04'
♅	05° ♑ 20'		-23°35'
♆	12° ♑ 22'		-21°57'
♇	14° ♏ 35'	℞	-00°38'
☊	04° ♓ 19'		-09°56'
Mc	08° ♓ 13'		-08°29'
As	00° ♊ 25'		+20°14'

Diagram Two
Sun / Venus conjunction
19th January, 1990

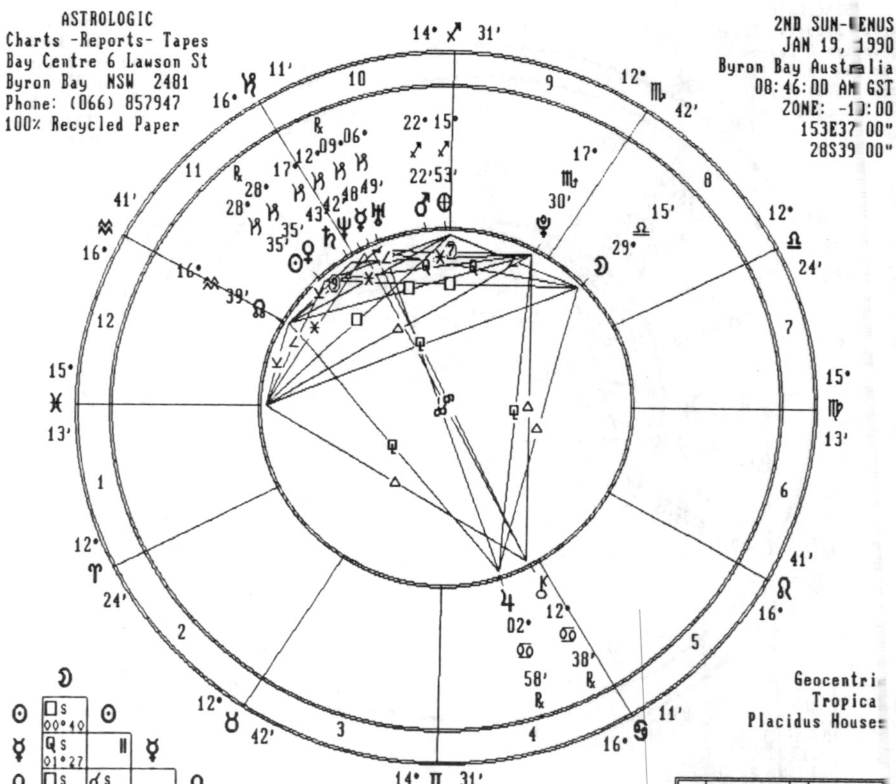

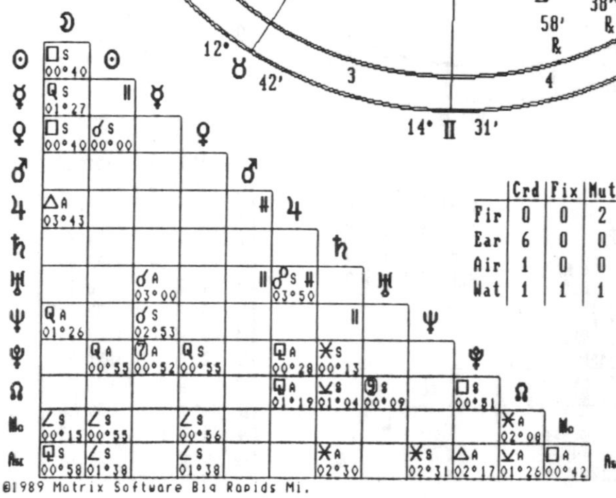

	Crd	Fix	Mut
Fir	0	0	2
Ear	6	0	0
Air	1	0	0
Wat	1	1	1

Pl	Geo Lon	℞	Declin.
☽	29° ♎ 15'		-15°54'
☉	28° ♑ 35'		-20°27'
☿	09° ♑ 48'	℞	-20°01'
♀	28° ♑ 35'	℞	-14°27'
♂	22° ♐ 22'		-23°27'
♃	02° ♋ 58'	℞	+23°20'
♄	17° ♑ 43'		-22°00'
♅	06° ♑ 49'		-23°32'
♆	12° ♑ 42'		-22°00'
♇	17° ♏ 30'		-02°14'
☊	16° ♒ 39'		-15°51'
Mc	14° ♐ 31'		-22°33'
As	15° ♓ 13'		-05°50'

Diagram Three
Sun / Venus conjunction
2nd Nov, 1990

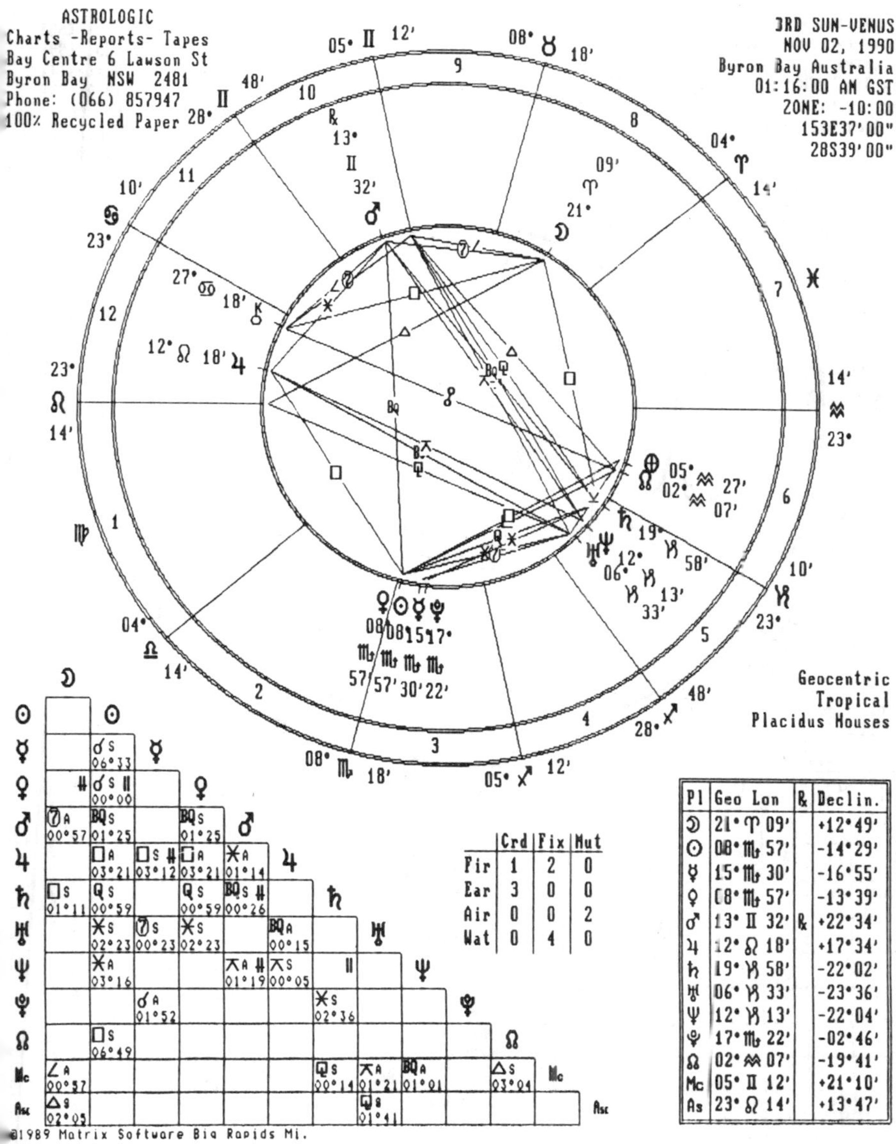

	Crd	Fix	Mut
Fir	1	2	0
Ear	3	0	0
Air	0	0	2
Wat	0	4	0

Pl	Geo Lon	℞	Declin.
☽	21° ♈ 09'		+12°49'
☉	08° ♏ 57'		-14°29'
☿	15° ♏ 30'		-16°55'
♀	08° ♏ 57'		-13°39'
♂	13° ♊ 32'	℞	+22°34'
♃	12° ♌ 18'		+17°34'
♄	19° ♑ 58'		-22°02'
♅	06° ♑ 33'		-23°36'
♆	12° ♑ 13'		-22°04'
♇	17° ♏ 22'		-02°46'
☊	02° ♒ 07'		-19°41'
Mc	05° ♊ 12'		+21°10'
As	23° ♌ 14'		+13°47'

Diagram Four
Sun / Venus conjunction
23rd August, 1991

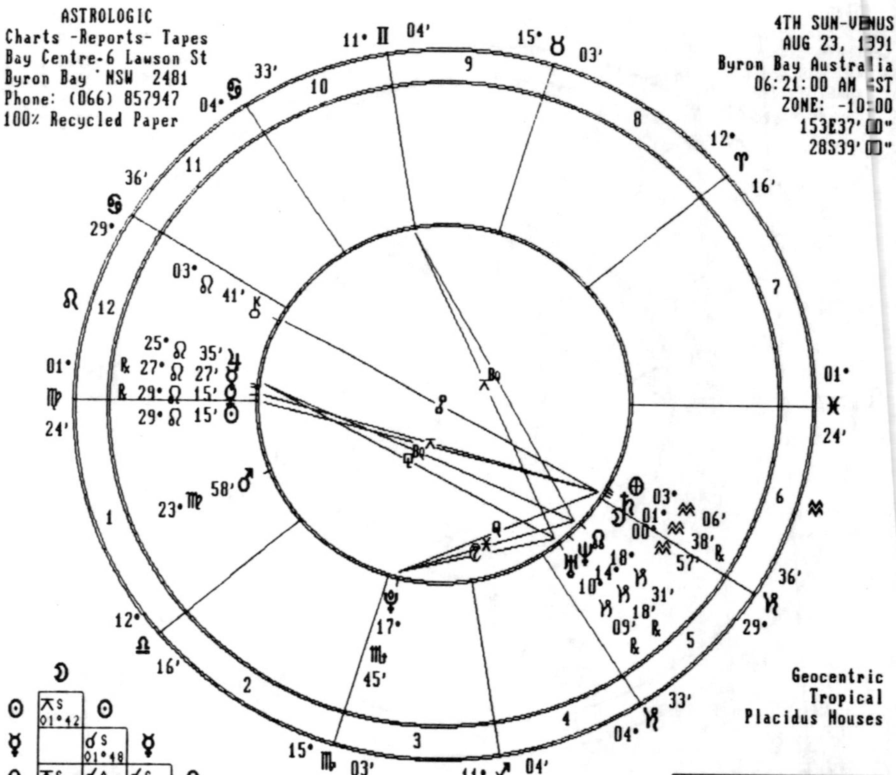

	☽	☉	☿	♀	♂	♃	♄	♅	♆	♇	☊	Mc
☉	⊼S 01°42											
☿		☌S 01°48										
♀	⊼S 01°42	☌A 00°00	☌S 01°48									
♂	△S 06°59			∥								
♃		☌S 03°39	☌A 01°52	☌A 03°39								
♄	☌A 00°41	⊼A 02°23										
♅			⚼A 02°18			⚼S 00°27						
♆		⚼A 00°03	⚼S 01°51	⚼S 00°03				☌S 04°09				
♇	QS 01°12				#		QA 01°53	⑦A 00°57				
☊						BQS 01°04			☌A 04°13	✶A 00°46		
Mc								⊼S 00°56			BQA 01°26	
Asc	⊼S 00°27	☌S 02°09	☌S 03°57	☌S 02°09			⊼A 00°14		⚼S 02°06		⚼A 02°07	

	Crd	Fix	Mut
Fir	0	4	0
Ear	2	0	2
Air	0	2	1
Wat	0	1	0

Pl	Geo Lon	℞	Declin.
☽	00° ♒ 57'		-18°51'
☉	29° ♌ 15'		+11°44'
☿	27° ♌ 27'	℞	+08°21'
♀	29° ♌ 15'	℞	+04°01'
♂	23° ♍ 58'		+03°07'
♃	25° ♌ 35'		+13°43'
♄	01° ♒ 38'	℞	-20°18'
♅	10° ♑ 09'	℞	-23°26'
♆	14° ♑ 18'	℞	-21°53'
♇	17° ♏ 45'		-02°46'
☊	18° ♑ 31'		-22°10'
Mc	11° ♊ 04'		+22°06'
As	01° ♍ 24'		+10°59'

Diagram Five
Sun / Venus conjunction
14th June, 1992

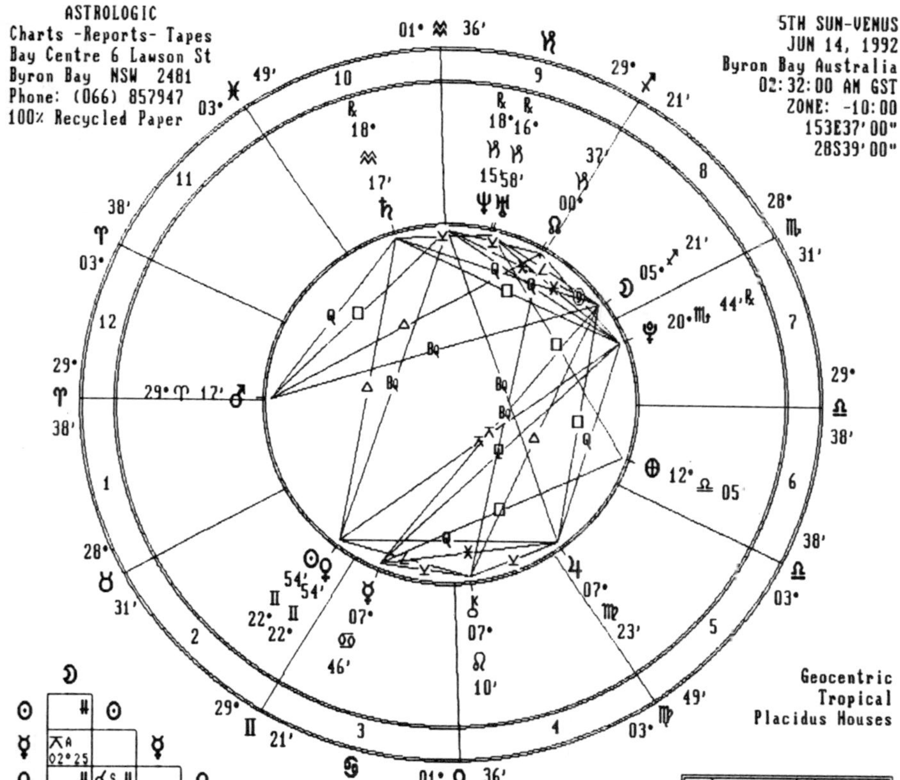

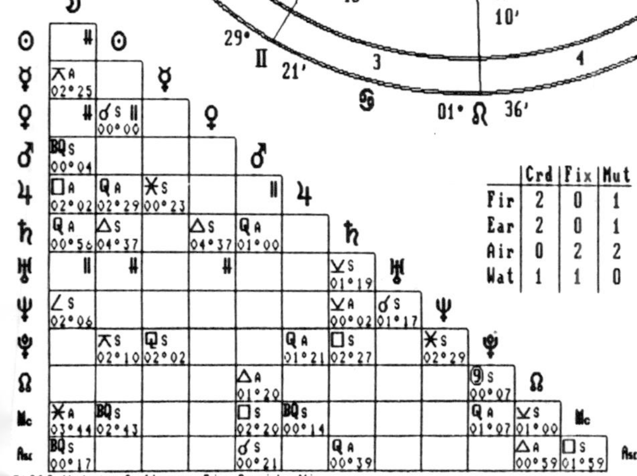

	Crd	Fix	Mut
Fir	2	0	1
Ear	2	0	1
Air	0	2	2
Wat	1	1	0

Pl	Geo Lon	℞	Declin.
☽	05° ♐ 21'		-23°25'
☉	22° ♊ 54'		+23°15'
☿	07° ♋ 46'		+25°11'
♀	22° ♊ 54'		+23°24'
♂	29° ♈ 17'		+10°04'
♃	07° ♍ 23'		+09°53'
♄	18° ♒ 17'	℞	-16°12'
♅	16° ♑ 58'	℞	-22°47'
♆	18° ♑ 15'	℞	-21°27'
♇	20° ♏ 44'	℞	-03°18'
☊	00° ♑ 37'		-23°26'
Mc	01° ♒ 36'		-19°48'
As	29° ♈ 38'		+11°21'

Diagram Six
Sun / Venus conjunction
1st April, 1993

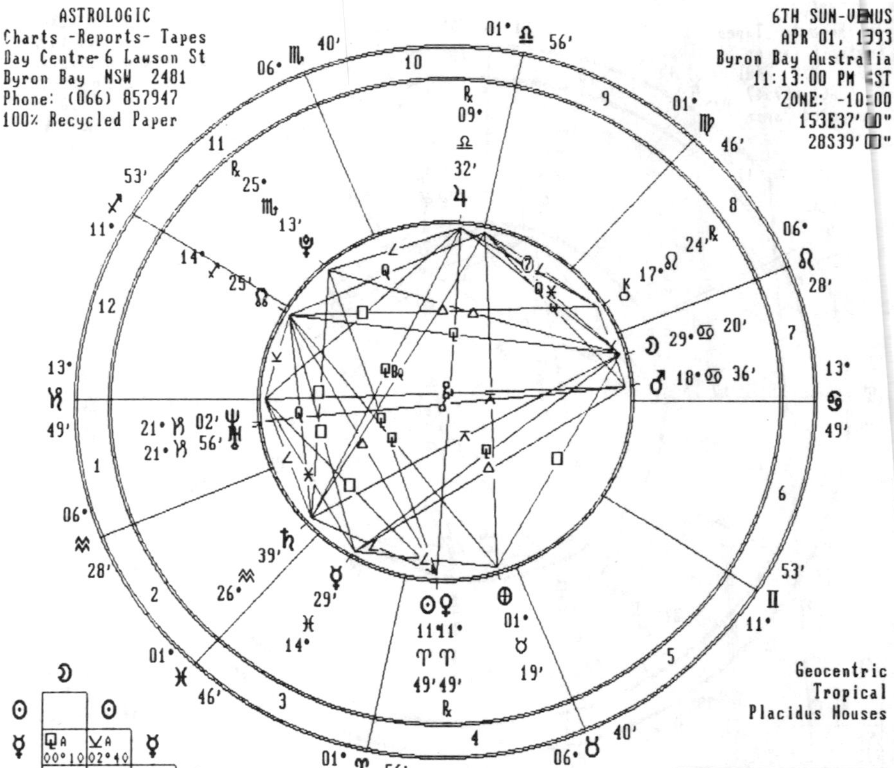

	☽	☉	☿	♀	♂	♃	♄	♅	♆	♇	☊	Mc
☉												
☿	⚼A 00°10	⚺A 02°40										
♀		☌S 00°00										
♂		□A 06°46	△A 04°06									
♃	QS 01°48	☍S 02°17		☍A 02°17								
♄	⚻S 02°40	∠S 00°10		∠A 00°10		⚼S 02°07						
♅					☍A 03°20							
♆					☍A 02°26			☌S ∥ 00°54				
♇	△S 04°07	⚼S ∥ 01°36		⚼A 01°36		∠S 00°41	□S 01°26					
☊	⚼A 00°05	△A 02°36	□S 00°05	△S 02°36			QS 00°14					
Mc	✱S 02°36				QS 01°20		BQA 00°43				QA 00°29	
Asc		□S 01°59	✱A 00°41	□S 02°00	☍A 04°47	□S 04°17	∠S 02°09				⚺A 00°36	

	Crd	Fix	Mut
Fir	2	0	0
Ear	3	0	0
Air	2	1	0
Wat	2	1	1

Pl	Geo Lon	℞	Declin.
☽	29° ♋ 20'		+16°46'
☉	11° ♈ 49'		+04°40'
☿	14° ♓ 29'		-07°13'
♀	11° ♈ 49'	℞	+11°54'
♂	18° ♋ 36'		+24°38'
♃	09° ♎ 32'	℞	-02°19'
♄	26° ♒ 39'		-13°41'
♅	21° ♑ 56'		-22°05'
♆	21° ♑ 02'		-21°08'
♇	25° ♏ 13'	℞	-04°43'
☊	14° ♐ 25'		-22°32'
Mc	01° ♎ 56'		-00°46'
As	13° ♑ 49'		-22°43'

Diagram Seven
Sun / Venus conjunction
17th January, 1994

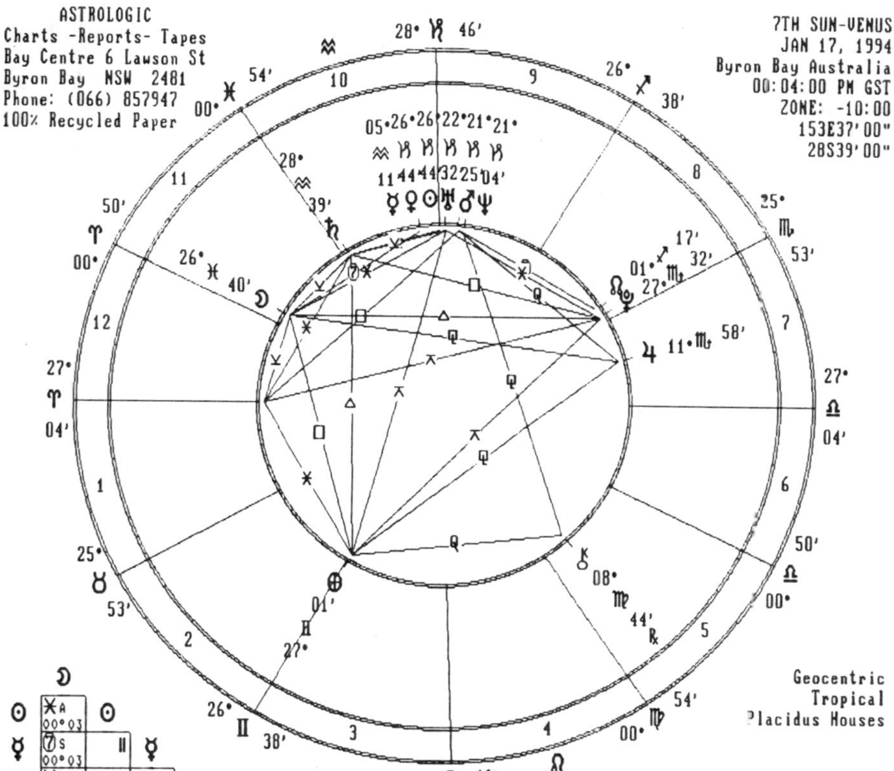

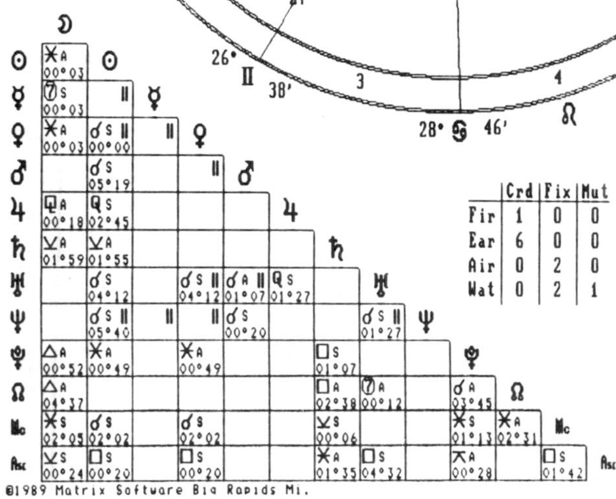

	Crd	Fix	Mut
Fir	1	0	0
Ear	6	0	0
Air	0	2	0
Wat	0	2	1

Pl	Geo Lon	℞	Declin.
☽	26° ♓ 40'		+02°55'
☉	26° ♑ 44'		-20°49'
☿	05° ♒ 11'		-20°58'
♀	26° ♑ 44'		-21°43'
♂	21° ♑ 25'		-22°39'
♃	11° ♏ 58'		-14°18'
♄	28° ♒ 39'		-13°16'
♅	22° ♑ 32'		-22°00'
♆	21° ♑ 04'		-21°11'
♇	27° ♏ 32'		-06°06'
☊	01° ♐ 17'		-20°25'
Mc	28° ♑ 46'		-20°24'
As	27° ♈ 04'		+10°26'

Diagram Eight
Sun / Venus conjunction
3rd November, 1994

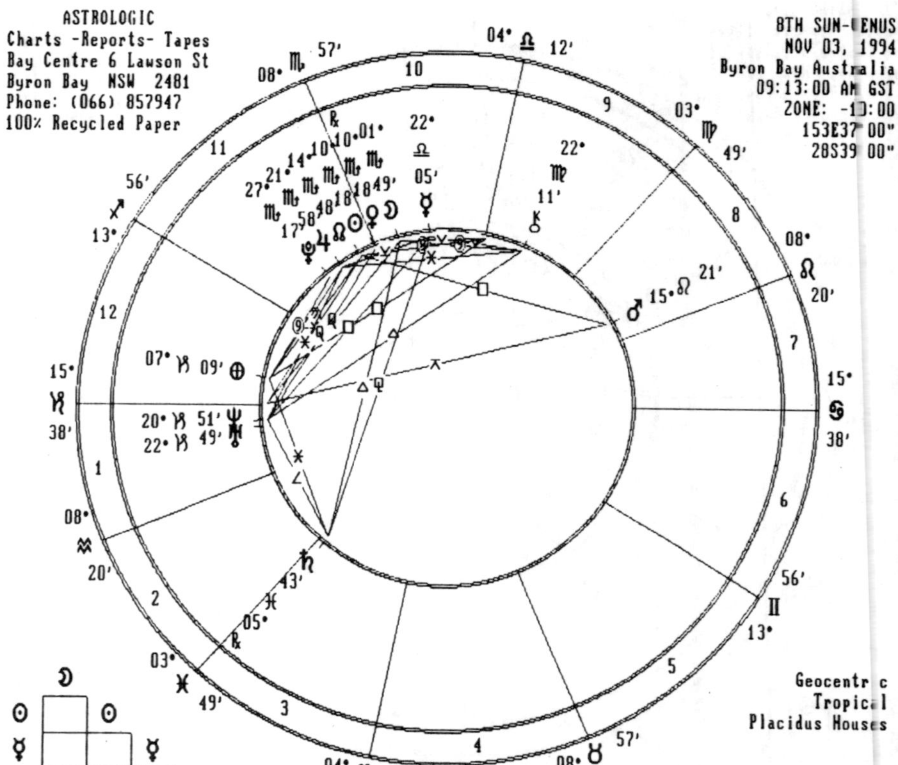

	☽	☉	☿	♀	♂	♃	♄	♅	♆	♇	☊	Mc
☉												
☿												
♀		☌S 00°00										
♂		□A 05°03										
♃			⊻S 00°07		‖							
♄	△A 03°53	△S 04°35	QS 01°22	△A 04°35								
♅		QA 00°31	□A 00°44	QS 00°31		✶A 00°52	∠S 02°06					
♆		QS 01°27	□S 01°14	QA 01°27		✶S 01°07	∠S 00°08	☌S ‖ 01°58				
♇			‖									
☊		☌A 04°30		☌S 04°30	□S 00°33							
Mc	⊻S 02°23										QA 00°36	
Asc	QS 01°48				⊼S 00°17						✶S 00°49	

	Crd	Fix	Mut
Fir	0	1	0
Ear	3	0	0
Air	2	0	0
Wat	0	5	1

Pl	Geo Lon	℞	Declin.
☽	01° ♏ 49'		-13°14'
☉	10° ♏ 18'		-14°55'
☿	22° ♎ 05'		-06°42'
♀	10° ♏ 18'	℞	-20°01'
♂	15° ♌ 21'		+17°47'
♃	21° ♏ 58'		-17°29'
♄	05° ♓ 43'	℞	-11°12'
♅	22° ♑ 49'		-22°00'
♆	20° ♑ 51'		-21°15'
♇	27° ♏ 17'		-06°35'
☊	14° ♏ 48'		-16°17'
Mc	04° ♎ 12'		-01°40'
As	15° ♑ 38'		-22°31'

Diagram Nine
Sun / Venus conjunction
21st August, 1995

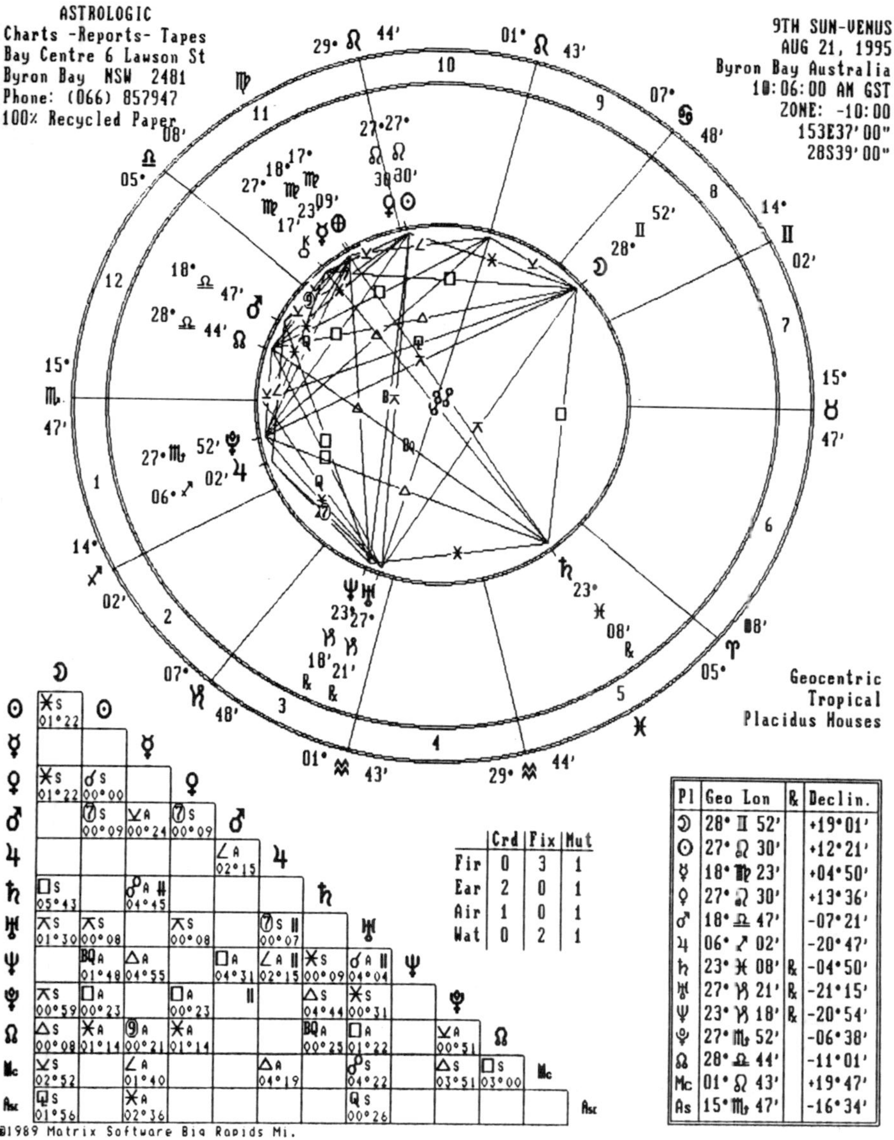

	☽	☉	☿	♀	♂	♃	♄	♅	♆	♇	☊	Mc
☉	✱S 01°22											
☿												
♀	✱S 01°22	☌S 00°00										
♂		⑦S 00°09	⊻A 00°24	⑦S 00°09								
♃					∠A 02°15							
♄	□S 05°43		☍A ∥ 04°45									
♅	⚻S 01°30	⚻S 00°08		⚻S 00°08		⑦S ∥ 00°07						
♆		BQA 01°48	△A 04°55		□A 04°31	∠A ∥ 02°15	✱S 00°09	☌A ∥ 04°04				
♇	⚻S 00°59	□A 00°23		□A 00°23	∥		△S 04°44	✱S 00°31				
☊	△S 00°08	✱A 01°14	⑨A 00°21	✱A 01°14			BQA 00°25	□A 01°22		⊻A 00°51		
Mc	⊻S 02°52		∠A 01°40			△A 04°19		☍S 04°22		△S 03°51	□S 03°00	
Asc	⚼S 01°56		✱A 02°36					⚼S 00°26				

	Crd	Fix	Mut
Fir	0	3	1
Ear	2	0	1
Air	1	0	1
Wat	0	2	1

Pl	Geo Lon	℞	Declin.
☽	28° ♊ 52'		+19°01'
☉	27° ♌ 30'		+12°21'
☿	18° ♍ 23'		+04°50'
♀	27° ♌ 30'		+13°36'
♂	18° ♎ 47'		-07°21'
♃	06° ♐ 02'		-20°47'
♄	23° ♓ 08'	℞	-04°50'
♅	27° ♑ 21'	℞	-21°15'
♆	23° ♑ 18'	℞	-20°54'
♇	27° ♏ 52'		-06°38'
☊	28° ♎ 44'		-11°01'
Mc	01° ♌ 43'		+19°47'
As	15° ♏ 47'		-16°34'

Diagram Ten
Sun / Venus conjunction
11th June, 1996

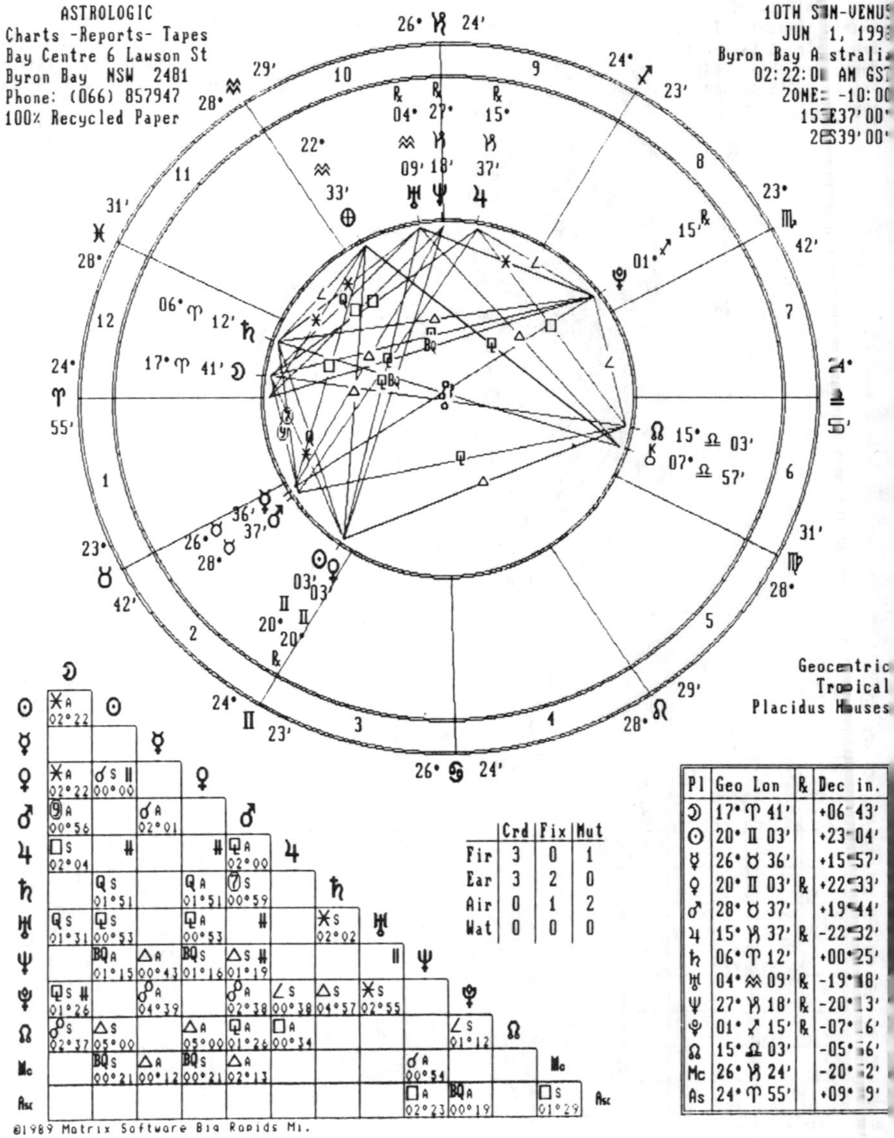

	☽	☉	☿	♀	♂	♃	♄	♅	♆	♇	☊	Mc
☉	⚹A 02°22											
☿												
♀	⚹A 02°22	☌S ∥ 00°00										
♂	⑨A 00°56		☌A 02°01									
♃	□S 02°04	#		#	⚼A 02°00							
♄		QS 01°51		QA 01°51	⑦S 00°59							
♅	QS 01°31	⚼S 00°53		⚼A 00°53	#		⚹S 02°02					
♆		BQA 01°15	△A 00°43	BQS 01°16	△S # 01°19			∥				
♇	⚼S # 01°26		☍A 04°39		☍A 02°38	∠S 00°38	△S 04°57	⚹S 02°55				
☊	☍S 02°37	△S 05°00		△A 05°00	⚼A 01°26	□A 00°34				∠S 01°12		
Mc		BQS 00°21	△A 00°12	BQS 00°21	△A 02°13				☌A 00°54			
Asc									□A 02°23	BQA 00°19		□S 01°29

©1989 Matrix Software Big Rapids MI.

	Crd	Fix	Mut
Fir	3	0	1
Ear	3	2	0
Air	0	1	2
Wat	0	0	0

Pl	Geo Lon	℞	Dec in.
☽	17° ♈ 41'		+06 43'
☉	20° ♊ 03'		+23 04'
☿	26° ♉ 36'		+15 57'
♀	20° ♊ 03'	℞	+22 33'
♂	28° ♉ 37'		+19 44'
♃	15° ♑ 37'	℞	-22 32'
♄	06° ♈ 12'		+00° 25'
♅	04° ♒ 09'	℞	-19° [illegible]8'
♆	27° ♑ 18'	℞	-20° [illegible]3'
♇	01° ♐ 15'	℞	-07° [illegible]6'
☊	15° ♎ 03'		-05° [illegible]6'
Mc	26° ♑ 24'		-20° [illegible]2'
As	24° ♈ 55'		+09° [illegible]9'

Diagram Eleven
Sun / Venus conjunction
2nd April, 1997

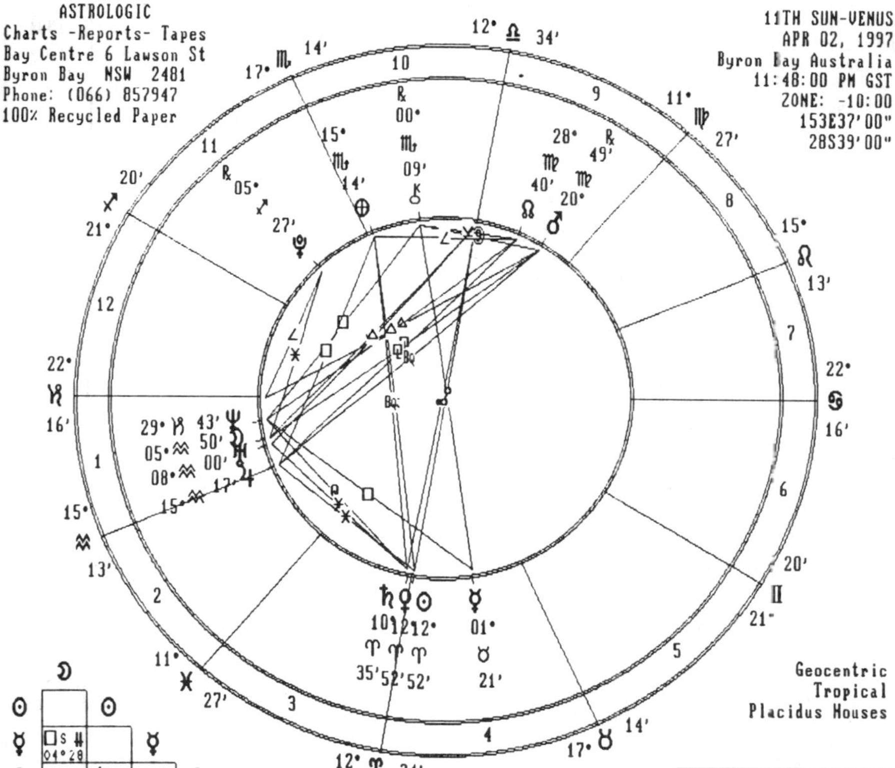

	☽	☉	☿	♀	♂	♃	♄	♅	♆	♇	☊	Mc
☉												
☿	□ S ∥ 04°28											
♀		☌ S 00°00										
♂	Q S 00°01											
♃		⚹ A 02°26		⚹ A 02°26	BQ S 00°29							
♄		☌ S 02°17		☌ S 02°17								
♅	☌ A 02°10				Q S 02°11		⚹ S 02°35					
♆	☌ S 06°06	Q S 01°08	□ S 01°38	Q S 01°08			Q A 01°09	∥				
♇	⚹ S 00°23							⚹ S 02°33				
☊						Q S 01°37			△ S 01°03			
Mc	△ S 06°44	☍ A 00°17		☍ A 00°17		△ A 02°43	☍ S 01°59	△ S 04°34				
Asc					△ S 01°28					∠ S 01°49		

	Crd	Fix	Mut
Fir	3	0	1
Ear	2	1	1
Air	1	3	0
Wat	0	0	0

Pl	Geo Lon	℞	Declin.
☽	05° ♒ 50'		-14°44'
☉	12° ♈ 52'		+05°05'
☿	01° ♉ 21'		+13°57'
♀	12° ♈ 52'		+03°54'
♂	20° ♍ 49'	℞	+06°36'
♃	15° ♒ 17'		-16°44'
♄	10° ♈ 35'		+02°10'
♅	08° ♒ 00'		-18°50'
♆	29° ♑ 43'		-19°48'
♇	05° ♐ 27'	℞	-08°33'
☊	28° ♍ 10'		+00°32'
Mc	12° ♎ 34'		-04°58'
As	22° ♑ 16'		-21°36'

Diagram Twelve
Venus / Neptune / Uranus conjunction
13th January, 1994

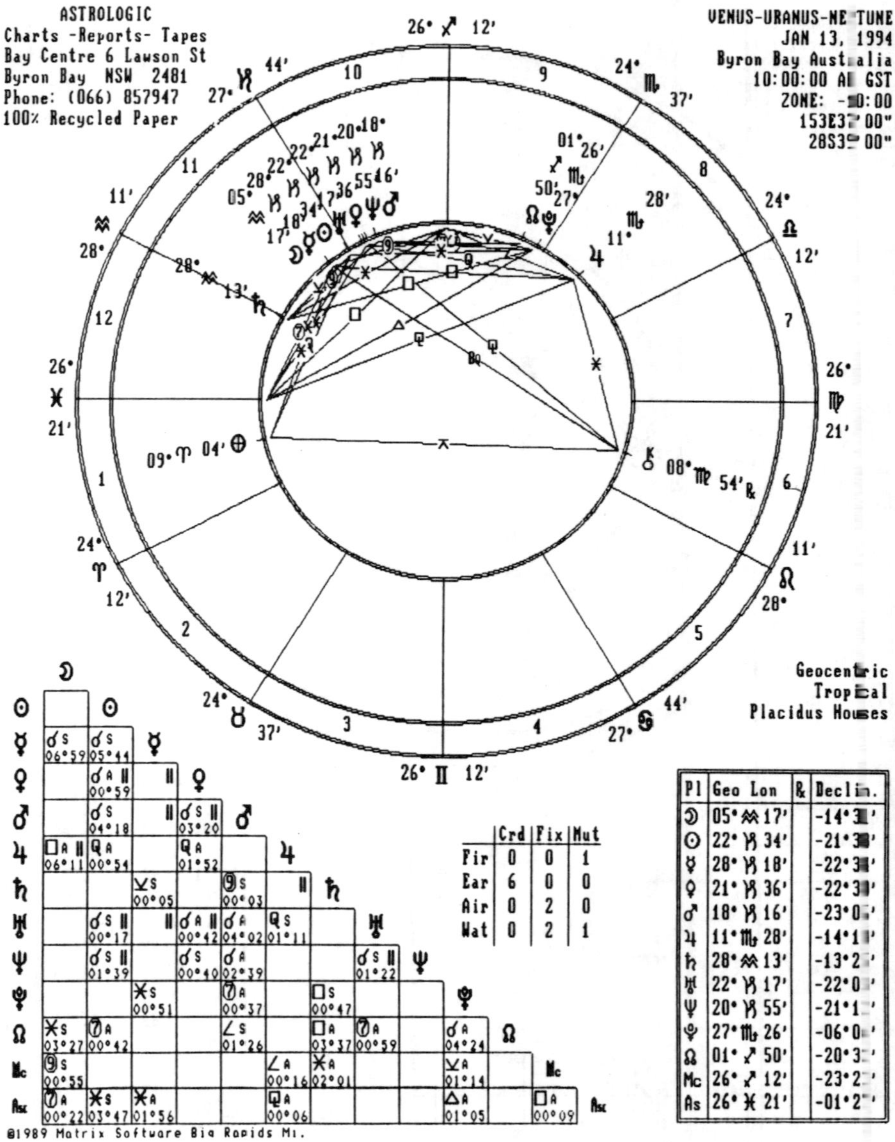

	☽	☉	☿	♀	♂	♃	♄	♅	♆	♇	☊	Mc
☉												
☿	☌ S 06°59	☌ S 05°44										
♀		☌ A ∥ 00°59	∥									
♂		☌ S 04°18	∥	☌ S ∥ 03°20								
♃	□ A ∥ 06°11	Q A 00°54		Q A 01°52								
♄			⊻ S 00°05		⑨ S 00°03	∥						
♅		☌ S ∥ 00°17	∥	☌ A ∥ 00°42	☌ A 04°02	Q S 01°11						
♆		☌ S ∥ 01°39		☌ S 00°40	☌ A 02°39			☌ S ∥ 01°22				
♇			✶ S 00°51		⑦ A 00°37		□ S 00°47					
☊	✶ S 03°27	⑦ A 00°42			∠ S 01°26		□ A 03°37	⑦ A 00°59		☌ A 04°24		
Mc	⑨ S 00°55					∠ A 00°16	✶ A 02°01			⊻ A 01°14		
Asc	⑦ A 00°22	✶ S 03°47	✶ A 01°56			Q A 00°06				△ A 01°05		□ A 00°09

©1989 Matrix Software Big Rapids MI.

	Crd	Fix	Mut
Fir	0	0	1
Ear	6	0	0
Air	0	2	0
Wat	0	2	1

Pl	Geo Lon	℞	Declin.
☽	05° ♒ 17'		-14°3[illegible]'
☉	22° ♑ 34'		-21°3[illegible]'
☿	28° ♑ 18'		-22°3[illegible]'
♀	21° ♑ 36'		-22°3[illegible]'
♂	18° ♑ 16'		-23°0[illegible]'
♃	11° ♏ 28'		-14°1[illegible]'
♄	28° ♒ 13'		-13°2[illegible]'
♅	22° ♑ 17'		-22°0[illegible]'
♆	20° ♑ 55'		-21°1[illegible]'
♇	27° ♏ 26'		-06°0[illegible]'
☊	01° ♐ 50'		-20°3[illegible]'
Mc	26° ♐ 12'		-23°2[illegible]'
As	26° ♓ 21'		-01°2[illegible]'

Appendix B

Attributes of Venus/Aphrodite

Numbers:	5 and 7
Colour:	Green
Gem:	Emerald
Metal:	Copper
Scent:	Rose, Frankinsense, Sandalwood
Flower:	Rose
Tree:	Myrtle
Fruit:	Pomegranate, quince, golden apples
Herb:	Clover, mint
Realm:	Nature's kingdom
Animals:	Lynx, sparrow, dove, swan, dolphin, serpent, goat
Force:	Magnetism
Quality:	Love, golden light
Tarot:	The Empress
Day:	Friday
Magical Weapons:	Lamp and girdle
Ritual:	Sacred marriage
Offspring:	Harmonia, Hermaphroditus, Rhodos, Eros, Priapus
Lovers:	Hephaestus, Ares, Hermes, Posieden
Flavours:	Sweet and warm
Forms:	Curved lines, rhythmic waves or scrolls
Musical Note:	A
Signs Ruled:	Libra and Taurus; Gemini (esoteric) and Capricorn (hierarchical)
Exalted In:	Pisces
In Detriment In:	Scorpio
Ray:	Fifth - Concrete Science
Key Words:	Love, beauty, magnetism, charm, harmony, peace, artistry, balance, culture, feminine, pleasure, partnership
Vices:	Indecision, lazy, indulgent, immoral

Bibliography

The Women's Spirituality Book, Diane Stein; Llewellyn Publications USA.

Woman's Mysteries, M Esther Harding; Rider Publications, London.

The Witches' Goddess, Janet & Stewart Farrar; Phoenix Publishing USA.

Mysteries of Venus, Mark Lerner; Great Bear Press USA.

Archetypes of the Zodiac, Kathleen Burt; Llewellyn Publications USA.

Goddess in Every Woman, Jean Shinoda Bolen; Harper and Row USA.

Esoteric Astrology, Alice Bailey; Lucis Trust USA.

The Labours of Hercules, Alice Bailey; Lucus Trust USA.

Treatise on Cosmic Fire, Alice A Bailey; Lucis Trust USA.

The Secret Doctrine, H P Blavatsky; Theosophical Publishing.

The Mayan Factor, Jose Arguelles; Bear & Co USA.

The Fixed Stars and Constellations in Astrology, Vivian E Robson; Samuel Weiser USA.

The Gnosis & the Law, Tellis S Papastavro.

The White Goddess, Robert Graves; Faber & Faber, London.

The Holy Bible, King James Version University Press, Oxford, England.

The I Ching, Richard Willhelm Translation Routledge & Kegan Paul, London.

The Mists of Avalon, Marion Bradley Sphere Books, London.

She - Understanding Feminine Psychology,
Robert A Johnson;
Harper & Row, USA

Androgyny - Towards a New Theory of Sexuality,
June Singer; Anchor Press USA.

The Astrologers Handbook,
Frances Sakoian & Louis S Acker,
Penguin, England.

The View over Atlantis,
John Michell;
Abacus, London.

Paracelsus - Life and Prophesies,
Franz Hartmann;
Rudolf Steiner Publications USA.

Healing Love through the Tao - Cultivating Female Sexual Energy, Mantak Chia & Maneewan Chia;
Healing Tao Books, New York, USA.

The Royal Road,
Stephan A Koeller;
Theosophical Publishing House USA.

The Prophet, Kahlil Gibran;
William Heinemann, London.

Reflections on the Christ, David Spangler;
Findhorn Foundation, Scotland.

The Great Cosmic Mother, Monica Sjoo & Barbara Mor;
Harper & Row USA.

Mothers Agenda, Mirra Alfasa

The Finding of the Third Eye,
Vera Stanley Alder; Rider & Co, London.

The Keys of Enoch,
J J Hurtak. Academy for Future Science, USA.

Secret Teachings of All Ages,
Manly P Hall
Philosophical Research Society USA.

Lord of the Rings,
J R Tolkien; Unwin Books, London.

The Harmonic Convergence and International Sacred Sites Festival of 1987,
Jim Berenholtz.

Index

A

B

C

D

E

I

J

K

L

M

N

O

P

Q

R

S

T

U

V